Biopolitics

An Advanced Introduction

Thomas Lemke

Translated by
Eric Frederick Trump

NEW YORK UNIVERSITY PRESS
New York and London

NEW YORK UNIVERSITY PRESS
New York and London
www.nyupress.org

References to Internet websites (URLs) were accurate at the time of writing.
Neither the author nor New York University Press is responsible for URLs
that may have expired or changed since the manuscript was prepared.

Library of Congress Cataloging-in-Publication Data
Lemke, Thomas.
[Biopolitik zur Einführung. English]
Biopolitics : an advanced introduction /
Thomas Lemke ; translated by Eric Frederick Trump.
p. cm. Includes bibliographical references and index.
ISBN 978-0-8147-5241-8 (cloth : alk. paper) — ISBN 978-0-8147-5242-5
(pbk. : alk. paper) — ISBN 978-0-8147-5299-9 (e-book)
1. Biopolitics. I. Title.
JA80.L46   2010
320.01—dc22   2010034537

New York University Press books are printed on acid-free paper,
and their binding materials are chosen for strength and durability.
We strive to use environmentally responsible suppliers and materials
to the greatest extent possible in publishing our books.

Manufactured in the United States of America

c 10 9 8 7 6 5 4 3 2 1
p 10 9 8 7 6 5 4 3 2 1

The translation of this work was supported by a grant from the
Goethe-Institut, which is funded by the German Ministry of Foreign Affairs.

Biopolitics

Contents

Foreword

IN CREATING THE "Biopolitics" book series for New York University Press, we hoped to achieve several intellectual and pragmatic goals. First, we wanted to solicit and encourage new book projects examining the potent intersection of medicine and technoscience with human bodies and lives. Second, we wanted to foster interdisciplinary scholarship in this field, realizing that contemporary "problems of the body" as they relate to technoscience and biomedicine can only be understood through diverse, overlapping, even competing analytical lenses. In this vein, the book series becomes a site for discourse about accounts of the body in relation to technologies, science, biomedicine, and clinical practices. Third, we were intent on encouraging scholarship in this field by established experts and emergent scholars. And finally, we were determined to offer fresh theoretical considerations of biopolitics alongside empirical and ethnographic work.

It is with regard to the goal of theoretical innovation that we are delighted to offer here the English translation of Thomas Lemke's *Biopolitics*, published originally in Germany. It is, of course, by now obvious that biopolitics, governmentality, and "life itself" have become concepts widely used in fields ranging from science and technology studies (STS) to biomedicalization studies, from cultural studies to security studies, from body/embodiment studies to health and illness studies. However, it is not the case that there has been substantial, or even adequate, theoretical conversation and debate about these terms and their usage. All too often, scholars take at face value the ideas of Foucault, for example, or Agamben as if it is quite clear to everyone what each has said and how their work might be applied to contemporary concerns. With the likely continued ascendance of scholarship on biopolitical processes and institutions, alongside investigation of their embodied consequences, we believe it is a timely task to engage in theoretical innovations regarding 21st-century biopolitics.

Thomas Lemke is an able guide through this biopolitical landscape. His research interests include social and political theory, organizational sociology, and social studies of genetics and reproductive technologies. He received his Ph.D. in political science in 1996 at the Johann Wolfgang Goethe University Frankfurt am Main, in Germany. He currently is Heisenberg Professor of Sociology with focus on Biotechnologies, Nature, and Society at the Faculty of Social Sciences of the Goethe University. From 1997 to 2006, he was an assistant professor of sociology at Wuppertal University. He also held visiting fellowships at Goldsmiths College in London (2001) and New York University (2003). A prolific scholar, Lemke has served on the editorial board of *Foucault Studies* and is currently an editor of *Distinktion: Scandinavian Journal of Social Theory*. He has published numerous journal articles in the areas of governmentality, risk, biopolitics, social theory, genetic technologies, and health and disease.

In *Biopolitics,* Lemke offers the first scholarly introduction to the idea of biopolitics. The book is, in his words, "a general orientation" designed to present a historical overview of the concept of biopolitics, while also exploring the term's relevance to contemporary theoretical conversations and debates. Yet at the same time, Lemke is quite reflexive about his project, recognizing that any such "systematic overview" necessarily represents the theoretical stance of its author. He contends that his is not a neutral account of the history of biopolitics as a social concept but rather a theoretical intervention in and of itself. Rather than solely a journey through theory's past, the book is a strategic intellectual intervention into the shifting and contested field of knowledge about biopolitics. As with our series of the same name, *Biopolitics* the book is *about* knowledge in the making at the same time that it *is* knowledge in the making. The book, then, is a volley in the ongoing conversation about what biopolitics is, how it relates to this thing called Life, and where we might go from here.

Lemke's ideas are broadly applicable. He gives us fresh ways of reading theorists such as Michel Foucault, Giorgio Agamben, Antonio Negri, Michael Hardt, Agnes Heller, Ferenc Fehér, Anthony Giddens, Didier Fassin, Paul Rabinow, and Nikolas Rose. He draws on geographically specific examples to illustrate his work, such as a discussion of Germany during World War II, but his theories are not located exclusively in his homeland—just as World War II was not specific to one nation. His final chapter is an exploration of some "neglected areas" of biopolitics, such as the work of Rudolf Goldscheid, vital politics, the Chicago School of human capital, and bioeconomics. Lemke's overview is not exhaustive, nor does he intend it to be. Rather, the book is meant to stimulate dialogue and to foster new scholarship on biopolitics. The absence of some theorists and ideas, such as feminist biopolitics, should not be read necessarily as deliberate elision but instead as an invitation to other scholars to produce future work in this area.

In short, Thomas Lemke's *Biopolitics* is significant in its engagement with a range of scholars, collecting in one book a valuable resource on biopolitical concepts, ideas, theorists, and origin stories. He has offered us who's who and what's what in this ever-expanding domain of knowledge. We look forward to the book's reception, as well as to the promise of other scholars using it as a springboard for further considerations.

Monica J. Casper and Lisa Jean Moore
SERIES EDITORS

Preface

THIS BOOK HAS emerged from a specific historical constellation. It addresses some crucial social and political events we have witnessed since the turn of the century. In the past ten years, intellectuals inside and outside the United States have used the notion of biopolitics to reflect on issues as heterogeneous as the war on terror after 9/11, the rise of neoliberalism, and biomedical and biotechnological innovations such as stem cell research, and the human genome project. In these debates, the concept of biopolitics has often served as an interpretive key to analyze how the production and protection of life is articulated with the proliferation of death; or it seeks to grasp how the reduction of human beings to "bare life" (e.g., in Guantánamo and Abu Ghraib) is linked to strategies to optimize and enhance human capabilities and life expectancy.

While many important political issues and theoretical questions have been addressed by employing the notion of biopolitics, it is often used in conflicting or even contradictory ways. However, the intensity of the debate and the prominence of biopolitics indicate that the term captures something essential in our present era. Nevertheless, there had until recently been no attempt to review the specific meanings of biopolitics in social theories and in philosophy. While biopolitics seemed to be everywhere, there was no attempt to contextualize and confront the different theoretical positions engaged in this debate. Against this background, I thought it might be useful to provide a systematic overview of the history of the notion of biopolitics and explore its relevance in contemporary theoretical debates.

The result of this intellectual experiment was originally published in 2007 in German, with the title *Biopolitik zur Einführung*. Although the present book is a translation of that volume, there are some significant changes to be noted. First, there is a new title: *Biopolitics: An Advanced Introduction*. As one of the anonymous reviewers of the book rightly pointed out, the original title, "Introduction to Biopolitics," might deter readers already familiar with the concept. In fact, the book offers more than an introduction. It identifies the historical dimensions of the notion of biopolitics and distinguishes systematically between conceptually different approaches. Second, I have also revised and updated the book. This edition includes literature on the topic published in the past two years and minor corrections and amendments. To make it more accessible for a U.S. readership, the present version limits references to German academic debates with which most readers would be unfamiliar and incorporates more literature in English.

I would like to express my gratitude to some individuals who made this edition possible, especially Eric Frederick Trump, who was responsible for the translation (I provided the translations of quotations from non-English sources), and Kevin Hall and Gerard Holden, who read and commented on the revisions I made to the original text. They all helped enormously to improve the quality of the book.

I am convinced that the book will find an interested readership among scholars and students in the United States and in the Anglophone world. It invites those working in areas as diverse as sociology, political science, cultural studies, anthropology, literature, legal studies, and philosophy to address questions that require us to go beyond neat disciplinary divisions of labor. But the book will most certainly also attract a larger audience already discussing the political impact of authors such as Michel Foucault, Giorgio Agamben, Michael Hardt, and Antonio Negri and those engaged in debates on the social and political implications of biotechnology and biomedicine. I hope that this small volume will contribute to the ongoing debate on biopolitics by providing the historical and theoretical knowledge to engage with the political issues at stake—and to define what politics means in biopolitical times.

Thomas Lemke
FRANKFURT AM MAIN
12 FEBRUARY 2010

Introduction

THE NOTION OF biopolitics has recently become a buzzword. A few years ago it was known only to a limited number of experts, but it is used today in many different disciplines and discourses. Beyond the limited domain of specialists, it is also attracting increasing interest among the general public. The term is used to discuss political asylum policies, as well as the prevention of AIDS and questions of demographic change. Biopolitics may refer to issues as diverse as financial support for agricultural products, promotion of medical research, legal regulations on abortion, and advance directives of patients specifying their preferences concerning life-prolonging measures.[1]

There is a range of diverse and often conflicting views about both the empirical object and the normative evaluation of biopolitics. Some argue strongly that "biopolitics" is necessarily bound to rational decision-making and the democratic organization of social life, while others link the term to eugenics and racism. The term figures prominently in texts of the Old Right, but it is also used by representatives of the New Left. It is used by both critics and advocates of biotechnological progress, by committed Marxists and unapologetic racists. A third line of disagreement concerns historical definitions and delimitations. Does biopolitics go back to antiquity or even to the advent of agriculture? Or, by contrast, is biopolitics the result of contemporary biotechnological innovations marking the beginning of a new era?

Plural and divergent meanings are undoubtedly evoked when people refer to biopolitics. This is surprising, since it is quite clear what the word literally signifies. It denotes a politics that deals with life (Greek: *bíos*). But this is where the problems start. What some people take to be a trivial fact ("Doesn't all politics deal with life?") marks a clear-cut criterion of exclusion for others. For the latter, politics is situated beyond biological life. From this point of view, "biopolitics" has to be considered an oxymoron, a combination of two contradictory terms. The advocates of this position claim that politics in the classical sense is about common action and decision-making and is exactly what transcends the necessities of bodily experience and biological facts and opens up the realm of freedom and human interaction.

This book seeks to bring clarity to this discussion by offering general orientation on the topic of biopolitics. Since this is the first introduction to this subject, I cannot rely on previous works or an established canon. Furthermore, biopolitics constitutes a theoretical and empirical field that crosses conventional disciplinary boundaries and undermines the traditional academic and intellectual division of labor. This introduction therefore has two objectives. On the one hand, it seeks to provide a systematic overview of the history of the notion of biopolitics; on the other hand, it explores its relevance in contemporary theoretical debates.

To avoid a possible misunderstanding, it should be made clear that this book does not intend to offer a neutral account or an objective representation of the diverse historical and contemporary meanings of "biopolitics." Defining biopolitics and determining its meaning is not a value-free activity that follows a universal logic of research. Rather, it is an integral part of a shifting and conflicting theoretical and political field. Each answer to the question of what processes and structures, what rationalities and technologies, what epochs and historical eras could be called "biopolitical" is always and inevitably the result of a selective perspective. In this respect, each definition of biopolitics must sharpen its analytical and critical profile against the blind spots and weak points of competing suggestions.

My point of departure is the virtual polarization that is attached to the merger of life and politics entailed in biopolitics. Existing understandings differ with respect to which part of the word they emphasize. It is possible to distinguish naturalistic concepts that take life as the basis of politics and to contrast these with politicist concepts, which conceive of life processes as the object of politics.[2] The former constitute a heterogeneous group of theories that I present in chapter 1. The spectrum runs from organicist concepts of the state in the first decades of the 20th century through racist modes of reasoning during National Socialism to biologistic ideas in contemporary political science. The politicist antipode configures biopolitics as a domain of practice or a subdiscipline of politics, aiming at the regulation and steering of life processes. Since the 1960s this line of interpretation has existed essentially in two different forms: first, as an ecological biopolitics that pursues conservative and defensive objectives and seeks to bind politics to the preservation and protection of the natural environment and, second, in a technical reading of biopolitics whose advocates are more interested in dynamic development and productivist expansion than in preservation and protection. The latter defines a new field of politics that is emerging as a result of new medical and scientific knowledge and biotechnological applications. This interpretation is especially popular nowadays, and is regularly cited in political discussions and media debates to describe the social and political implications and potential of biotechnological innovations. I present the different dimensions of the politicist discourse in chapter 2.

The central thesis of the book is that both lines of interpretation fail to capture essential dimensions of biopolitical processes. Apart from their obvious differences, the politicist and the naturalist position share some basic assumptions. Both conceptions are based on the idea of a stable hierarchy and an external relationship between life and politics. The advocates of naturalism regard life as being "beneath" politics, directing and explaining political reasoning and action. The politicist conception sees politics as being "above" life processes; here, politics is more than "pure" biology, going beyond the necessities of natural existence. Each fundamental position on the problem of biopolitics relies on the stability of one pole of the semantic field in order to explain variations in the other pole. Either biology accounts for politics, or politics regulates biology. However, this means that both conceptions fail to explain the instability and fragility of the border between "life" and "politics"—and it is exactly this instability that has prompted so many people to employ the notion of biopolitics. Since the two approaches take "life" and "politics" as isolated phenomena, they are both unable to account for their relationality and historicity. The emergence of the notion of biopolitics signals a double negation (cf. Nancy 2002): in contrast to naturalist positions, life does not represent a stable ontological and normative point of reference. The impact of biotechnological innovations has demonstrated that life processes are transformable and controllable to an increasing degree, which renders obsolete any idea of an intact nature untouched by human action. Thus, nature can only be regarded as part of nature-society associations. At the same time, it has become clear that biopolitics also marks a significant transformation of politics. Life is not only the object of politics and external to political decision-making; it affects the core of politics—the political subject. Biopolitics is not the expression of a sovereign will but aims at the administration and regulation of life processes on the level of populations. It focuses on living beings rather than on legal subjects—or, to be more precise, it deals with legal subjects that are at the same time living beings.

Against the naturalist and the politicist reading I propose a relational and historical notion of biopolitics that was first developed by the French philosopher and historian Michel Foucault. According to Foucault, life denotes neither the basis nor the object of politics.

Instead, it presents a border to politics—a border that should be simultaneously respected and overcome, one that seems to be both natural and given but also artificial and transformable. "Biopolitics" in Foucault's work signals a break in the order of politics: "the entry of phenomena peculiar to the life of the human species into the order of knowledge and power, into the sphere of political techniques" (1980, 141–142). Foucault's concept of biopolitics assumes the dissociation and abstraction of life from its concrete physical bearers. The objects of biopolitics are not singular human beings but their biological features measured and aggregated on the level of populations. This procedure makes it possible to define norms, establish standards, and determine average values. As a result, "life" has become an independent, objective, and measurable factor, as well as a collective reality that can be epistemologically and practically separated from concrete living beings and the singularity of individual experience.

From this perspective, the notion of biopolitics refers to the emergence of a specific political knowledge and new disciplines such as statistics, demography, epidemiology, and biology. These disciplines make it possible to analyze processes of life on the level of populations and to "govern" individuals and collectives by practices of correction, exclusion, normalization, disciplining, therapeutics, and optimization. Foucault stresses that in the context of a government of living beings, nature does not represent an autonomous domain that has to be respected by governmental action but depends on the practices of government itself. Nature is not a material substratum to which practices of government are applied but the permanent correlative of those practices. The ambivalent political figure "population" plays a decisive role in this process. On the one hand, population represents a collective reality that is not dependent on political intervention but is characterized by its own dynamics and modes of self-regulation; this autonomy, on the other hand, does not imply an absolute limit to political intervention but is, on the contrary, the privileged reference of those interventions.

The discovery of a "nature" of the population (e.g., rates of birth and death, diseases, etc.) that might be influenced by specific incentives and measures is the precondition for directing and managing it. Chapter 3 discusses the different dimensions of the notion of biopolitics in the work of Foucault. In the following chapters, I present lines of reception and correctives emanating from Foucault's concept of biopolitics.

Giorgio Agamben's writings and the works of Michael Hardt and Antonio Negri are certainly the most prominent contributions to a reformulation of Foucault's notion of biopolitics. Their respective theories assign a strategic role to demarcation and delimitation. According to Agamben, it is the basic separation of "bare life"—the form of existence reduced to biological functions—and political existence that has shaped Western political history since antiquity. He argues that the constitution of sovereign power requires the production of a biopolitical body and that the institutionalization of law is inseparably connected to the exposure of "bare life." Hardt and Negri diagnose a new stage of capitalism that is characterized by the dissolution of the boundaries between economy and politics, production and reproduction. Whereas Agamben criticizes Foucault for neglecting the fact that modern biopolitics rests on a solid basis of a premodern sovereign power, Hardt and Negri hold that Foucault did not recognize the transformation of modern into postmodern biopolitics. Their respective contributions to the discussion are analyzed in chapters 4 and 5.

The following chapters examine two main lines of reception that have taken up Foucault's work on biopolitics. The first focuses on the *mode of politics* and asks how biopolitics is to be distinguished historically and analytically from "classical" forms of political representation and articulation. In chapter 6, I concentrate on a discussion of the works of Agnes Heller and Ferenc Fehér, who observe a regression of politics resulting from the increasing significance of biopolitical issues. Then I present Anthony Giddens's concept of life politics (which does not explicitly refer to Foucault) and Didier Fassin's idea of biolegitimacy.

The second strand of thought focuses on the *substance of life*. Scholars working along these lines ask how the foundations, means, and objectives of biopolitical interventions have been transformed by a biotechnologically enhanced access to the processes of life and the human body. Looking more closely at the work of these scholars in chapter 7, I discuss concepts of molecular politics, thanatopolitics, and anthropopolitics and the ideas of "biosociality" (Paul Rabinow) and "ethopolitics" (Nikolas Rose).

Chapter 8 is devoted to a neglected area of biopolitics. It presents a series of theoretical concepts which suggest that biopolitics cannot be separated from the economization of life. The approaches covered include the idea of an "economy of humans" (*Menschenökonomie*) developed by the Austrian social theorist and sociologist Rudolf Goldscheid at the beginning of the 20th century. This is followed by the concept of a "vital politics" as promoted by German liberals after World War II and the theory of human capital developed by the Chicago School. The final section focuses on visions of a "bioeconomy" in contemporary political action plans and some recent empirical studies that critically evaluate the relations between biotechnological innovations and transformations in capitalism. Chapter 9 integrates the diverse refinements of and amendments to the Foucauldian notion of biopolitics into an "analytics of biopolitics." I seek to demonstrate the theoretical importance of this research perspective. Finally, I show how this analytical framework differs from bioethical discourse.

If these sometimes quite heterogeneous chapters have become a whole and if the result is a "lively" introduction (meaning a vivid and comprehensive presentation) to the field of biopolitics, this is due to a number of readers and colleagues who have helped me with their suggestions and comments. I received important ideas and valuable criticism from Martin Saar, Ulrich Bröckling, Robin Celikates,

Susanne Krasmann, Wolfgang Menz, Peter Wehling, Caroline Prassel, and Heidi Schmitz. Ina Walter assisted me with the technical work on the text, and Steffen Herrmann attentively read and corrected the manuscript. The constructive discussions at the Institute for Social Research in Frankfurt helped to sharpen my arguments. Finally, I would like to thank the German Research Foundation for funding work on the book by awarding me a Heisenberg Grant.

1

Life as the Basis of Politics

State Biology: From Organicist Concepts to Racist Concepts

Although the concept of biopolitics has now become familiar, it may not be widely known that it has nearly a hundred-year history. Its initial appearance was as part of a general historical and theoretical constellation. By the second half of the 19th century, *Lebensphiloso-phie* (the philosophy of life) had already emerged as an independent philosophical tendency; its founders were Arthur Schopenhauer and Friedrich Nietzsche in Germany and Henri Bergson in France. The individual *Lebensphilosophen* (philosophers of life) represented quite diverse theoretical positions. They shared, however, the reevaluation of "life" and its adoption as a fundamental category and normative criterion of the healthy, the good, and the true. Life—understood as bodily fact or organic existence, as instinct, intuition, feeling, or "experience" (*Erlebnis*)—was opposed to the "dead" and the "petrified," which were represented by the "abstract" concept, "cold" logic, or the soulless "spirit." The concept of life served as a standard by which processes perceived as adversarial to life, such as processes of rationalization, civilization, mechanization, and technologization, were subjected to critical examination.

The concept of biopolitics emerged in this intellectual setting at the beginning of the 20th century. The Swedish political scientist Rudolf Kjellén may have been among the first to employ it.[1] Kjellén, until his death in 1922 a professor at the University of Uppsala, had an organicist concept of the state and considered states "super-individual creatures . . . , which are just as real as individuals, only disproportionately bigger and more powerful in the course of their development" (1924, 35). For Kjellén, the natural form of statehood is the nation-state, which expresses the state's "ethnic individuality" (ibid., 103). The "state as form of life" is ultimately characterized, in his view, by social struggles over interests and ideas articulated by classes and groups. In conjunction with this conviction, Kjellén introduces the concept of biopolitics: "In view of this tension typical of life itself . . . the inclination arose in me to baptize this discipline after the special science of biology as biopolitics; . . . in the civil war between social groups one recognizes all too clearly the ruthlessness of the life struggle for existence and growth, while at the same time one can detect within the groups a powerful cooperation for the purposes of existence" (1920, 93–94).

Kjellén was not alone in understanding the state as a "living organism" or a "living creature." Many of his contemporaries—political scientists and specialists in public law, as well as biologists and health professionals—conceived of the state as a collective subject that ruled over its own body and spirit. Many of these people saw in politics, economics, culture, and law merely expressions of the same organic powers, which constitute the state and determine its specific characteristics (cf. Selety 1918; Uexküll 1920; Hertwig 1922; Roberts 1938). The organicist concept understands the state not as a legal construction whose unity and coherence is the result of individuals' acts of free will but as an original form of life, which precedes individuals and collectives and provides the institutional foundation for their activities. The basic assumption is that all social, political, and legal bonds rest on a living whole, which embodies the genuine and the eternal, the healthy, and the valuable. The reference to "life" serves here both as a mythic starting point and as a normative guideline. Furthermore, it eludes every rational foundation or democratic decision-making. From this perspective, only a politics that orients itself toward biological laws and takes them as a guideline can count as legitimate and commensurate with reality.

During the period of National Socialism the antidemocratic, conservative character of the organicist concept of the state acquired a racist bias. The widely used metaphor of "the people's body" (*Volkskörper*) at this time designated an authoritarian, hierarchically structured, and racially homogeneous community. There were two central features of the National Socialist conception of state and society. First, it promoted the idea that the subjects of history were not individuals, groups, or classes but self-enclosed communities with a common genetic heritage. This idea was complemented by the assumption of a natural hierarchy of peoples and races according to their different "inherited biological quality," such that it seemed not only justified but also necessary to treat individuals and collectives unequally. Second, National Socialist ideology rested on the belief that social relations and political problems could ultimately be attributed to biological causes. At the same time, representatives of the regime regularly denied concepts of biological determinism and stressed that natural, organic facts were essentially "historical and spiritual" facts. As a result, education and willpower were regarded as having a decisive meaning for the development of individuals and collectives. In the words of the well-known geneticist Otmar von Verschuer, "Hereditary predisposition means the possibility of reaction. Environment determines which of the given possibilities is realized" (1936, 10).

The National Socialist concept of biopolitics is marked by the constitutive tension between, on the one hand, the idea of life as a fateful power and the site of mythical origin and, on the other hand, the conviction that active modification and control of biological events is possible. To formulate and elaborate its social and political conception of itself, the National Socialist movement made use of many different sources, integrating social Darwinist ideas along with Pan-Germanic and nationalist ideologies. It took up anthropological, biological, and medical concepts and simultaneously stimulated the production of theories and empirical work in these disciplines (see

Weingart, Kroll, and Bayertz 1992). Since heterogeneous ideas frequently stand alongside one another in National Socialist texts, it is difficult to speak of a coherent conception of biopolitics. Here I focus only on two general characteristics that decisively marked National Socialist biopolitics: first, the foundation of the biopolitical program in racial hygiene and "hereditary biology" (*Erbbiologie*) and, second, the combination of these ideas with geopolitical considerations.

Hans Reiter, the president of the Reich Health Department, explained the racial underpinning of "our biopolitics" in a speech in 1934. This speech demonstrated that the representatives of National Socialism regarded biopolitics as a break with classical concepts of politics. Reiter claimed that the past, present, and future of each nation was determined by "hereditary biological" facts. This insight, he said, established the basis for a "new world of thinking" that had developed "beyond the political idea to a previously unknown world view" (1939, 38). The result of this understanding was a new, biologically grounded concept of people and state: "It is inevitable that this course of thought should lead to the recognition of biological thinking as the baseline, direction, and substructure of every effective politics" (ibid.). The goal of this policy consisted of improving the German people's "efficiency in living" (*Lebenstüchtigkeit*) by a quantitative increase of the population and a qualitative improvement in the "genetic materials" of the German people. In order to achieve this, Reiter recommended negative and positive eugenic practices. Accordingly, inferior offspring were to be avoided, while the regime supported all those who were regarded as "biologically valuable" (ibid., 41). However, National Socialist biopolitics comprised more than "selection" and "elimination." Laws, regulations, and policies governing racial politics had as their objective not only the regulation and disciplining of reproductive behavior; they also contained responses to the imaginary dangers of "racial mixing." The development and maintenance of genetic material was, in this light, only possible through protection against the "penetration of foreign blood" and the preservation of the "racial character" of the German people (ibid., 39). Concerns about the purity of the "race" coincided with the battle against internal and external national enemies. At this point, biopolitical ideas join with geopolitical considerations. The combination of the racial political program with the doctrine of *Lebensraum* (living space) provided the ideological foundation for the imperialist expansion of the Nazi Reich.

The concept of *Lebensraum*, which was by 1938 at the latest a central element of National Socialist foreign policy, goes back to scientific ideas that had been worked out earlier in the 20th century. The "father" of geopolitics was the German geographer Friedrich Ratzel, who coined the word *Lebensraum* around the turn of the century. His "anthropogeography" examined the relationship between the motionless Earth and the movements of peoples, in which two geographical factors play a central role: space and position. Kjellén was also familiar with the concept of geopolitics and used it in his political writings.

The most important figure in German geopolitics, however, was Karl Haushofer, who occupied a chair in geography at the University of Munich. Haushofer was Rudolf Hess's teacher and friend and contributed substantially to the founding of the *Zeitschrift für Geopolitik* (Journal for Geopolitics), the first volume of which appeared in 1924 (Neumann 1942, 115–124). In one of the issues of this journal, an author named Louis von Kohl explained that biopolitics and geopolitics were together "the basis for a natural science of the state" (1933, 306). This "biology of the state," as envisioned by Kohl, examined the development of a people or a state from two different but complementary points of view: "When we observe a people or a state we can place greater emphasis either on temporal or spatial observations. Respectively, we will have to speak of either biopolitics or geopolitics. Biopolitics is thus concerned with historical development in time, geopolitics with actual distribution in space or with the actual interplay between people and space" (ibid., 308).

Kohl distinguishes between a vertical and a horizontal perspective on society and state. The first envisages the development of the people's body and its "living space" in time. It concentrates on "the importance of racial elements" and observes "the swelling and ebbing of the people's body, the social stratifications it consists of and their changes, its susceptibility to sicknesses, and so forth" (ibid., 308). This viewpoint corresponds to a horizontal perspective that tries to comprehend the struggles and conflicts of "different powers and fields of power in geographical space" (ibid., 309). Temporal development and spatial movement should be considered together. They serve Kohl as a guideline and yardstick for politics.

The link between racial delusion and genocide contained in the formula "*Blut und Boden*" (blood and soil) may have been a peculiarity of National Socialist biopolitics. The fundamental idea of a "biologization of politics" is nevertheless neither a German idiosyncrasy nor limited to the period of National Socialism. The state's "gardening-breeding-surgical ambitions" (Baumann 1991, 32) can be traced back at least to the 18th century. In the period between World War I and World War II, these fantasies blossomed in ideologically and politically antagonistic camps. They emerged in the projects of the "new Soviet man" under Stalin's dictatorship but also in the eugenic practices of liberal democracies. German racial hygienists were in close scientific contact with geneticists around the world and turned to American sterilization programs and practices of immigration restriction to promote their own political positions (Kevles 1995). Like the Nazi regime, Stalinist ideologues sought to use new scientific knowledge and technological options to "refine" and "ennoble" the Soviet people. Biopolitical visions not only crossed national boundaries; they were also supported by nonstate actors and social movements. The Rockefeller Foundation, which played a significant role in funding the rise of molecular biology in the United States in the 1930s, expected this science to produce new knowledge and instruments of social control and to be able to steer and to optimize human behavior (Kay 1993).

Even if racist biopolitics no longer had any serious scientific or political standing after the end of the Third Reich and the atrocities of World War II, it continued to have appeal. Representatives of right-wing movements still use the concept of biopolitics today, in order to complain about the ignorance of the *"Zeitgeist"* toward the "question of race"; they contend that the category of race has continuing relevance for the present. Like the National Socialist ideologues, they diagnose a fundamental social crisis resulting from the alleged struggle between different "races" and the imagined threat of "racial mixing" and "degeneration." One example of this persistent theme is a book by Jacques Mahieu, formerly a member of the Waffen SS, who fled to Argentina after World War II and taught political science there in various universities. In order to establish a "foundation for politics," the author believes political science's "important role" today consists in defining the causes of the increasing "racial struggles" and "ethnic collisions" (2003, 13). Beyond representing a model to specify the problem, the biopolitical triad of People-Nation-Race evoked in the title of Mahieu's book is also meant to offer solutions to the crisis it claims to identify. "The meaning of biopolitics" is, according to the author, "to calculate the totality of genetic processes insofar as they influence the life of human communities" (ibid., 12).

Biopolitology: Human Nature and Political Action

In the middle of the 1960s a new theoretical approach developed within political science which advanced a "naturalistic study of politics" (Blank and Hines 2001, 2). "Biopoliticians" (Somit and Peterson 1987, 108) use biological concepts and research methods in order to investigate the causes and forms of political behavior.[2] Representatives of this approach draw on ethological, genetic, physiological, psychopharmacological, and sociobiological hypotheses, models, and findings. Despite research and publication activity that now spans four decades, it is only in the United States that one can find a rudimentary institutionalization of this theoretical perspective today. The Association for Politics and Life Sciences (APLS) acquired an official section of the American Political Science Association (APSA) in 1985 but lost it ten years later because of declining membership. The journal founded by this section, *Politics and the Life Sciences*, has been in existence since 1982 (Blank and Hines 2001, 6–8). Outside the United States, this branch of political science plays hardly any role, even if there are scholars in a few countries who consider themselves biopoliticians.[3]

Even among advocates for this approach, however, its meaning and scope are disputed. Whereas some biopoliticians demand a paradigm shift in political science or want to integrate all the social sciences into a new, unified sociobiological science (Wilson 1998), others see in this approach an important supplement to and perfection of already established theoretical models and research methods. Within this heterogeneous field of research, it is possible to identify four areas to which most of the projects can be assigned. The first area comprises reception of neo-Darwinian evolutionary theory. At its center stands the historical and anthropological question of the development of human beings and the origins of state and society. A second group of works takes up ethological and sociobiological concepts and findings in order to analyze political behavior. Works interested in physiological factors and their possible contribution to an understanding of political action fall into the third category. A fourth group focuses on practical political problems ("biopolicies"), which arise from interventions in human nature and changes to the environment (Somit and Peterson 1987, 108; Kamps and Watts 1998, 17–18; Blank and Hines 2001; Meyer-Emerick 2007).

Despite the diversity of the theoretical sources and thematic interests involved here, one can nonetheless speak of a common research perspective since most of these works agree on three fundamental aspects. First, the object of investigation is primarily political behavior, which—and this is the underlying assumption—is caused in a substantial way by objectively demonstrable biological factors. Within these explicative models, (inter)subjective motivations or reasons play no more than a minor role, as do cultural factors. Second, the objective of the approach is not the interpretation of symbolic structures or the provision of normative critique; it is much more oriented toward describing and explaining observable behavior in order to draw conclusions for a rational politics, that is, a politics consistent with biological exigencies. Third, methodologically speaking, the approach rests on the perspective of an external observer who objectively describes certain forms of behavior and institutional processes. By contrast, concepts that approach reality from the perspective of actors or participants are considered scientifically deficient (Saretzki 1990, 86–87).

Common to all representatives of "biopolitics" is thus a critique of the theoretical and methodological orientation of the social sciences, which, in their view, is insufficient. They argue that the social sciences are guided by the assumption that human beings are, in principle, free beings, a view that gives too much significance to processes of learning and socialization and thereby fails to see that human (political) behavior is in large part biologically conditioned. From this perspective, the "culturalism" of the social sciences remains "superficial" as it systematically ignores the "deeper" causes of human behavior. Conventional social-scientific research is thus "one-sided" and "reductionist" insofar as the biological origins of human behavior remain outside the horizon of the questions it poses. In order to produce a "more realistic" evaluation of human beings and how they live, biopoliticians demand a "biocultural" or "biosocial" approach. This is supposed to integrate social-scientific and biological viewpoints, in order to replace a one-sided either-or with a combinatory model (Wiegele 1979; Masters 2001; Alford and Hibbing 2008).

Biopoliticians do not as a rule assume a deterministic relationship but refer to biological "origins" or "factors" which are supposed to decisively shape the motives and spaces of political actors. They presume that in human evolutionary history a multitude of behavioral patterns have arisen and that although none of these completely determines human behavior, many mold it to a considerable degree in various areas of life. Works written under the rubric of "biopolitics" are interested above all in competition and cooperation, anxiety and aggression, relations of dominance, the construction of hierarchies, enmity toward foreigners, and nepotism. These phenomena ultimately go back—or at least this is the assumption—to evolutionary mechanisms and lead to the formation of affects that usually guide individuals in the direction of "biologically beneficial" behavior. According to this view, the formation and persistence of states depend less on democratic consensus or social authority than on psychologically and physically grounded relationships of dominance, which can in turn be traced back to inherited behavior patterns (cf. Wiegele 1979; Blank and Hines 2001).

In this view, the emergence of hierarchies in human society is not a social phenomenon but rather an inevitable result of evolutionary history. The reason given for this is that asymmetrically distributed opportunities for access and participation allegedly offer evolutionary advantages, since stable and predictable relationships are supposed to favor the transmission of one's genes to the next generation. In order to establish solid grounds for this assumption, biopoliticians often present economic propositions and premises as matters of natural fact. Accordingly, human beings are by nature disposed to competition over scarce resources, and insofar as they are differently equipped biologically for competitive situations, power is distributed unequally. For this reason, social hierarchies are said to be necessary and unavoidable (Somit and Peterson 1997).

Furthermore, preferences for certain forms of government and authority are derived from human evolutionary history. It is regularly assumed that the genetic endowment of human beings makes authoritarian regimes likelier than democratic states. A democratic state is, according to this view, only possible under particular—and very rarely occurring—evolutionary conditions. A democracy can only arise and assert itself against the dominating behavior of individuals and groups if power resources are distributed widely enough so that no actor can achieve supremacy (Vanhanen 1984). Even ethnocentrism and ethnic conflict are traced back to determinants in human phylogeny, to conflict over scarce resources and the principle of kin selection. The latter idea assumes that in smaller groups the welfare of the group member is more highly valued than the welfare of nonmembers, due to a higher probability of being biologically related to one another (Kamps and Watts 1998, 22–23).

Taken together, the works of biopoliticians reveal a rather pessimistic image of human beings and society. Nonetheless, it would be wrong to equate "biopolitics" across the board with National Socialist or racist positions. No one particular political orientation follows necessarily if one assumes the existence of inborn characteristics. In fact, the political positions of biopoliticians vary considerably. The spectrum extends from avowed social reformers such as Heiner Flohr (1986) to authors whose arguments follow distinctively racist patterns, for example J. Philippe Rushton, who traces the higher prevalence of criminality among African Americans in the United States to inherited behavior related to skin color (1998). To analyze the approach with the tools of ideological critique is not sufficient. The thesis *that* biological factors play a role in the analysis of social and political behavior is not the problem; the question is, rather, *how* the interaction is understood—and in this respect the responses of the biopoliticians are not at all convincing. A long list of reservations and objections has been put forward in response to the research perspectives they suggest. In the following I briefly present some of them.

Although biopoliticians programmatically demand that biological knowledge should be taken into account in the social sciences, how exactly "biological" factors on the one hand and "cultural" and "social" factors on the other interact, and how they should be delineated against one another, are issues that remain largely unexplained in their work. Furthermore, it is unclear how the alleged "biological basis" concretely "evokes" or "produces" particular patterns of political behavior. The one-dimensional concept of genetic regulation promoted by many representatives of this approach (e.g., the idea of genes "for" hierarchy or dominant behavior) no longer corresponds to current findings in biological science and has been increasingly criticized in recent years (Oyama, Griffiths, and Gray 2001; Neumann-Held and Rehmann-Sutter 2006). In general, there is no systematic consideration of the manner in which diverse scientific cultures could be conceptually, theoretically, and methodologically integrated. As a result, the claim to have provided "deeper" empirical explanations and the promise of a more comprehensive theoretical and conceptual approach remain largely unfounded and unrealized (Saretzki 1990, 91–92). In starting from the idea that "nature" is an autonomous system and a closed sphere, with the conviction that this closed sphere decisively shapes political action, biopoliticians put forward and prolong the very dualism of nature and society whose continuing existence they also bemoan.

A further problem with the "biopolitical" approach is that representatives of this type of research pay too little attention to symbolic structures and cultural patterns of meaning for the investigation of political processes. Thus, by only treating social phenomena from the perspective of their alignment with natural conditions, they grasp little of what they claim to study. They are not sensitive to the question of how far sociopolitical evolution affects and changes "biological factors." Biopoliticians therefore see "the human being" as a product of biocultural processes of development only, not as a producer of these processes. This one-sided perspective conceals a crucial dimension in the present discussion of the relationship between nature and society, biology and politics:

At a moment when, with the development of new genetic and reproductive technologies, the capacity has also increased to selec-
tively or even constructively shape one's own biological evolution
in totally new dimensions, the point is no longer to become aware
of putatively neglected "biological conditions." By now, these have
become contingent in a completely new way. When a society can
discuss the "fabrication of nature" and "human beings made to
measure," first and foremost the question of the goals of and responsibility for the shaping of nature more and more strongly by
society becomes important—as does an institutional design in
whose framework these new contingencies can be adequately dealt
with. (Saretzki 1990, 110–111; cf. also Esposito 2008, 23–24)

This very question, the question of institutional and political forms
and the social answers to the "question of nature," provides the point
of departure for the second line of inquiry addressing "biopolitics."

2

Life as an Object of Politics

Ecological Biopolitics

In the 1960s and early 1970s, the meaning of biopolitics assumed another form. It was not so much focused on the biological foundations of politics but rather disclosed life processes as a new object of political reflection and action. In light of the ecological crisis that was increasingly being addressed by political activists and social movements, biopolitics now came to signify policies and regulatory efforts aimed at finding solutions to the global environmental crisis. These efforts received an important stimulus from the Report to the Club of Rome (Meadows et al. 1972), which demonstrated through scientific modeling and computer simulations the demographic and ecological limits of economic growth. The report demanded political intervention to halt the destruction of the natural environment. Along with growing awareness of the limits of natural resources and anxiety about the consequences of a "population explosion," apocalyptic scenarios also multiplied. It was postulated that nothing less than life on the planet and the survival of the human species were at stake.

In this context, the concept of biopolitics acquired a new meaning. It came to stand for the development of a new field of politics and political action directed at the preservation of the natural environment of humanity. This was clear, for example, in the six-volume series *Politik zwischen Macht und Recht* (Politics between Power and Law) by the German political scientist Dietrich Gunst, who, in addition to writing about the German constitution and foreign policy, also dedicated a volume to biopolitics. According to Gunst, biopolitics embraces "anything to do with health policy and the regulation of the population, together with environmental protection and questions concerning the future of humanity. This political arena in its comprehensive form is comparatively new and takes into consideration the fact that questions about life and survival are increasingly relevant" (1978, 9).

The individual chapters of the book focus on the political and social problems that result from a growing world population, starvation and difficulties securing proper nutrition in many countries, air and water pollution, the depletion of natural resources, and dwindling energy supplies. The organization of health care, biomedical innovations, and the "manipulation of life and death" (ibid., 21) play only a marginal role in the book. After an overview of the fields of action and the political challenges they pose, Gunst comes to the general conclusion that these worsening problems will be solved only through a "life-oriented politics" (ibid., 12). What the author means by this phrase are those measures and initiatives that would help to achieve an ecological world order. It will be necessary, he believes, to align economic structures (consumption, production, distribution, etc.), as well as political activities at local, regional, national, and international levels, with biological exigencies (ibid., 165–183).

The concept of biopolitics was linked to ecological considerations and became a reference point for various ideological, political, and religious interests. One of the most curious responses to the "ecological question" is the idea of a "Christian biopolitics" put forward by theologian Kenneth Cauthen in his book *Christian Biopolitics: A Credo and Strategy for the Future* (1971). The author asserts the emergence of a "planetary society," which comes into existence once the biological frontiers of Earth are exceeded. The book explores the dangers arising from and the opportunity for a fundamental change in consciousness that would be caused by such a development. According to Cauthen, a transformation in ideas, goals, and attitudes is necessary in order to bring about the desired transition, and this is where theology and the church have a special role to play. "Christian biopolitics" consists in developing "a religio-ethical perspective centered on life and the quest for enjoyment in a science-based technological age. This ecological model requires an organic understanding of reality. Such an understanding interprets man as a biospiritual unity whose life is set within cosmic nature, as well as within human history" (Cauthen 1971, 11–12). More specifically, Cauthen aims at promoting "a movement toward an ecologically optimum world community full of justice and joy in which the human race can not only survive but embark on exciting new adventures of physical and spiritual enjoyment" (ibid., 10).

However, authors motivated by religious beliefs were not the only ones to use emerging environmental debates for their own ends. Many representatives of right-wing movements were especially active in Germany and united the ecological message with eugenic and racist motifs. As early as 1960, the German division of the *Weltbund zum Schutze des Lebens* (World Union for Protection of Life) was founded, and the *Gesamtdeutsche Rat für Biopolitik* (All-German Council on Biopolitics) was established five years later. In 1965, a supplement to the German right-wing magazine *Nation Europa* appeared with the title *Biopolitik*. Contributors to this issue concentrated on "two undesirable biopolitical trends": the "wildly advancing overpopulation of the Earth" and the "mixing together of all races and genealogical lines," which leads to a "sullying of the gene pool" (*Nation Europa* 1965, 1). The contributors claimed that in order to preserve "life's possibilities for our children," the politics of the future must be biopolitics, and its goal must be the eradication of these two fundamental problems facing humanity (ibid., 1-2). However, at stake was not only the "care of the genetic health of future generations" (ibid., 45) and the control of the world's population. Right-wing groups were also, relatively early, very engaged in the struggle against "nuclear death" and health problems resulting from nuclear energy (cf., for example, *Biologische Zukunft* 1978).

Technocentric Biopolitics

The idea of biopolitics as securing and protecting global natural foundations of life was soon augmented by a second component. The 1970s were not only the decade in which a growing environmental movement and enhanced sensitivity to ecological questions emerged; these years also saw several spectacular biotechnological innovations. In 1973, it was possible for the first time to transfer DNA from one species to another. With this accomplishment, genetic information from different organisms could be isolated and recombined in various ways. Around the same time, the diagnosis of fetuses became an integral part of prenatal care, and new reproductive technologies such as in-vitro fertilization were developed.

The growing significance of genetic and reproductive technologies raised concerns about the regulation and control of scientific progress. If the results of biological and medical research and its practical applications demonstrated how contingent and fragile the boundary between nature and culture is, then this intensified political and legal efforts to reestablish that boundary. It was deemed necessary to regulate which processes and procedures were acceptable and under what conditions. There was also a need to clarify what kind of research would be supported with public funding and what would be prohibited.

Such questions led ultimately to a second stratum of meaning in biopolitics, one that is situated close to the considerations and concerns of bioethics. These relate to the collective negotiation of, and agreement on, the social acceptability of what is technologically possible. The German sociologist Wolfgang van den Daele provides an exemplary definition of this strand of biopolitics. He writes that biopolitics refers to

> the approximately twenty-year societal thematization and regulation of the application of modern technologies and natural science

to human life. Within the purview of these policies stand, above all, reproductive medicine and human genetics. Increasingly, however, one finds brain research, as well as the scientifically and technically rather uninteresting field of cosmetic surgery. Biopolitics responds to the transgression of boundaries. It reacts to the fact that the boundary conditions of human life, which until now were unquestioned because they lay beyond the reach of our technical capabilities, are becoming accessible to us. . . . The results of such transgressions are moral controversies and debates about regulation that come down to the old question: Just because we can, should we? (2005, 8)

In recent years, this interpretation has become dominant in journalism and in political declarations and speeches. Since at least the turn of the millennium, biopolitics has stood for administrative and legal procedures that determine the foundations and boundaries of biotechnological interventions.[1]

It is safe to say, then, that since the 1970s "life" has become a reference point for political thinking and political action in two respects. On the one hand, we can say that the human "environment" is threatened by the existing social and economic structures and that policymakers need to find the right answers to the ecological question and to secure the conditions of life on Earth and the survival of humanity. On the other hand, it is becoming increasingly difficult to know, because of bioscientific discoveries and technological innovations, what exactly the "natural foundations" of life are and how these can be distinguished from "artificial" forms of life. With the transformation of biology into a practice of engineering, and the possibility of perceiving living organisms not as self-contained and delimited bodies but rather as constructs composed of heterogeneous and exchangable elements (e.g., organs, tissues, DNA), traditional environmental protection and species conservation efforts are becoming less pertinent. This is the case because their self-understanding is still rooted in the assumption of separate orders of being and of the existence of nature as a domain that is in principle free from human intervention. In this light, Walter Truett Anderson notes a shift from "environmentalism to biopolitics" (1987, 94). The latter represents a new political field that gives rise to hitherto unforeseen questions and problems and that goes well beyond the traditional forms of environmental protection. As Anderson sees it, biopolitics not only comprises measures to save endangered species but also should tackle the problem of "genetic erosion" and regulate biotechnological progress (ibid., 94–147).

As a result of this problematization, the ecological version of biopolitics was weakened until ultimately it was integrated into the technocentric variant. If the former assigned itself a task that tended toward the conservative and defensive, pursuing the goal of preserving natural foundations of life, the latter is more dynamic and productivist, concerned with the exploitation of these foundations. The ecological interpretation of biopolitics was in this respect locked into a naturalistic logic, as it strove to thematize the interaction between natural and societal processes and so to determine the correct political answers to environmental questions. Central to the technocentric version of biopolitics, however, is not the adaptation of "society" to a separate "natural environment" but rather the environment's modification and transformation through scientific and technological means.

Of course, these interpretive threads are difficult to tease apart historically or systematically. Thus, for example, "green" genetic technology is regularly promoted with the dubious argument that it solves central environmental and development policy problems. If nothing else, the synthesis of the ecological and technocentric strands of biopolitics represents a programmatic promise that strives to inspire hope for a world in which the means of production will be energy efficient, low in pollution, and protective of natural resources, a world which has overcome hunger through an increase in food production (for a critical appraisal of this view, see, e.g., Shiva and Moser 1995).

The German philosopher Volker Gerhardt puts forward a comprehensive definition of biopolitics that encompasses both the ecological and technocentric approaches. Gerhardt sees biopolitics as a "wide-ranging domain of action" characterized by "three main tasks." Along with "ecologically securing the basics of life" and "the biological increase of the benefits of life," the protection of the development of life through medical intervention has also become an issue (2004, 32). The challenges posed by the last domain have radically changed and expanded the range of contemporary biopolitics. Gerhardt writes that it now includes "those questions in which the human becomes an object of the life sciences" (ibid., 44). He laments the broad range of skepticism and reprimand that stretches from representatives of the Church to Marxists. These people put "biopolitics under general suspicion" (ibid., 37) and foment irrational fears about new technologies.

In the face of such critics, Gerhardt demands as a political duty a rational debate about the possibilities and risks of technology. According to him, it is necessary to have a political culture that respects the freedom of the individual and takes care to ensure that the human being remains an end in itself (ibid., 30):

> Since biopolitics to a certain degree impinges upon our self-understanding as human beings, we must insist on its link to basic liberties and to human rights. And since it can have wide-ranging consequences for our individual self-understanding, it also makes demands on the individual conduct of our lives. If one does not wish biotechnology to interfere with questions which are situated within the discreet sanctuary of love, one must make this decision first and above all for oneself. (Ibid., 36)

This appeal fails to convince. This is because, on the one hand, basic liberties and human rights are hardly suited to complement or correct biotechnological innovations, since the right to life takes a central position in most constitutional and legal texts. When the Universal Declaration of Human Rights guarantees "life, liberty, and security of person" in Article 3, and many national constitutions grant special protection to the life and health of their citizens, these guarantees are not so much a limitation of biotechnological options as a way of broadening them. On the other hand, the allusion to autonomous decision-making processes and individual choice is on the whole too limited, in that the conditions under which these processes occur might call forth new constraints. As prenatal diagnosis shows, the very possibility of prebirth examinations forces the couple in question to take a decision, namely, whether to make use of the diagnostic option. Moreover, the decision against prenatal diagnosis is still a decision and not comparable to the state of ignorance before such diagnostic methods were available. Should a child with physical or mental disabilities be born, the parents could be held responsible for their decision not to use prenatal diagnostics and selective abortion.

The key question, which neither Gerhardt nor other representatives of the politicist version of biopolitics answer, is the question of the "we" that is regularly engaged with in these debates. Who is it who decides about the contents of biopolitics and decides autonomously on one's conduct of life? The interpretation of biopolitics as a mere province of traditional politics is inadequate, in that it presumes that the substance of the political sphere remains untouched by the growing technological possibilities for regulating life processes. This, however, is not the case. Biopolitical questions are fundamental precisely because not only are they objects of political discourse, but they also encompass the political subject him- or herself. Should embryonic stem cells be considered legal subjects or biological material? Does neurobiological research reveal the limits of human free will? In such cases, the question is not just about the political assessment of technologies or the negotiation of a political compromise in a field of competing interests and value systems. Rather, the question is who should participate in such decision-making and evaluative processes and how normative concepts of individual freedom and responsibility interact with biological factors. In this respect, biopolitics defines

> the borderland in which the distinction between life and action is *introduced* and dramatized in the first place. This distinction is nothing less than a constitutive element of politics as the development of the citizen's will and decision-making powers. Biopolitics is in this respect not a new, ancillary field of politics, but rather a problem space at the heart of politics itself. (Thomä 2002, 102; emphasis in original)

Biopolitics cannot simply be labeled a specific political activity or a subfield of politics that deals with the regulation and governance of life processes. Rather, the meaning of biopolitics lies in its ability to make visible the always contingent, always precarious difference between politics and life, culture and nature, between the realm of the intangible and unquestioned, on the one hand, and the sphere of moral and legal action, on the other.

It is not enough, then, to dissolve these distinctions in one direction or another, either as a way of promoting a stronger delimitation of politics and its adaptation to biological conditions or in order to celebrate the broadening of the political field—a field that encompasses sets of problems that were once understood as natural and self-evident facts but that are now open to technological or scientific intervention.

The notion of biopolitics calls into question the topology of the political. According to the traditional hierarchy, the political is defined as humanity elevating itself as *zoon politikon* above mere biological existence. Biopolitics shows that the apparently stable boundary between the natural and the political, which both naturalist and politicist approaches must presuppose, is less an origin than an effect of political action. When life itself becomes an object of politics, this has consequences for the foundations, tools, and goals of political action. No one saw more clearly this shift in the nature of politics more clearly than Michel Foucault.

3

The Government of Living Beings: Michel Foucault

IN THE 1970s, the French historian and philosopher Michel Foucault introduced a concept of biopolitics that broke with the naturalist and politicist interpretations that were discussed in the preceding chapters. In contrast to the former conception of biopolitics, Foucault describes biopolitics as an explicit rupture with the attempt to trace political processes and structures back to biological determinants. By contrast, he analyzes the historical process by which "life" emerges as the center of political strategies. Instead of assuming foundational and ahistorical laws of politics, he diagnoses a historical break, a discontinuity in political practice. From this perspective, biopolitics denotes a specific modern form of exercising power.

Foucault's concept of biopolitics orients itself not only against the idea of processes of life as a foundation of politics. It also maintains a critical distance from theories that view life as the object of politics. According to Foucault, biopolitics does not supplement traditional political competencies and structures through new domains and questions. It does not produce an extension of politics but rather transforms its core, in that it reformulates concepts of political sovereignty and subjugates them to new forms of political knowledge. Biopolitics stands for a constellation in which modern human and natural sciences and the normative concepts that emerge from them structure political action and determine its goals. For this reason, biopolitics for Foucault has nothing to do with the ecological crisis or an increasing sensibility for environmental issues; nor could it be reduced to the development of new technologies. Rather, biopolitics stands for a fundamental transformation in the order of politics:

> For the first time in history . . . biological existence was reflected in political existence. . . . But what might be called a society's "threshold of modernity" has been reached when the life of the species is wagered on its own political strategies. For millennia, man remained what he was for Aristotle: a living animal with the additional capacity for a political existence; modern man is an animal whose politics places his existence as a living being in question. (Foucault 1980, 142–143)

Foucault's use of the term "biopolitics" is not consistent and constantly shifts meaning in his texts. However, it is possible to discern three different ways in which he employs the notion in his work. First, biopolitics stands for a historical rupture in political thinking and practice that is characterized by a rearticulation of sovereign power. Second, Foucault assigns to biopolitical mechanisms a central role in the rise of modern racism. A third meaning of the concept refers to a distinctive art of government that historically emerges with liberal forms of social regulation and individual self-governance. But it is not only the semantic displacements that are confusing. Foucault not only employs the term "biopolitics"; he also sometimes uses the word "biopower," without neatly distinguishing the two notions. I briefly discuss the three dimensions of biopolitics in this chapter before addressing the role of resistance in the context of biopolitical struggles.

Making Live and Letting Die

Although the notion of biopolitics appeared for the first time in Foucault's work in a lecture he gave in 1974 (2000a, 137), it is systematically introduced only in 1976 in his lectures at the Collège de France and in the book *The History of Sexuality, Vol. 1* (Foucault 2003 and

1980, respectively). In this work, Foucault undertakes an analytical and historical delimitation of various mechanisms of power while contrasting sovereign power with "biopower." According to him, the former is characterized by power relations operating in the form of "deduction": as deprivation of goods, products, and services. The unique character of this technology of power consists in the fact that it could in extreme cases also dispose of the lives of the subjects. Although this sovereign "right of life and death" only existed in a rudimentary form and with considerable qualification, it nevertheless symbolized the extreme point of a form of power that essentially operated as a right to seizure. In Foucault's reading, this ancient right over death has undergone a profound transformation since the 17th century. More and more it is complemented by a new form of power that seeks to administer, secure, develop, and foster life:

> "Deduction" has tended to be no longer the major form of power but merely one element among others, working to incite, reinforce, control, monitor, optimize, and organize the forces under it: a power bent on generating forces, making them grow, and ordering them, rather than one dedicated to impeding them, making them submit, or destroying them. (Foucault 1980, 136)

The integration of sovereign power into biopower is by no means a transformation within politics alone. Rather, it is itself the result of some important historical transformations. Decisive for the "entry of life into history" (ibid., 141) was the increase of industrial and agricultural production in the 18th century, as well as growing medical and scientific knowledge about the human body. Whereas the "pressure exerted by the biological on the historical" (ibid, 142) in the form of epidemics, disease, and famine was quite high until that time, the technological, scientific, social, and medical innovations allowed now for a "relative control over life. . . . In the space for movement thus conquered, and broadening and organizing that space, methods of power and knowledge assumed responsibility for the life processes and undertook to control and modify them" (ibid., 142).

Foucault sees the particularity of this biopower in the fact that it fosters life or disallows it to the point of death, whereas the sovereign power takes life or lets live (2003, 241). Repressive power over death is subordinated to a power over life that deals with living beings rather than with legal subjects. Foucault distinguishes "two basic forms" of this power over life: the disciplining of the individual body and the regulatory control of the population (1980, 139). The disciplinary technology to supervise and control the individual body had already emerged in the 17th century. This "anatomo-politics of the human body" (ibid.) conceives of the human body as a complex machine. Rather than repressing or concealing, it works by constituting and structuring perceptual grids and physical routines. In contrast to more traditional forms of domination such as slavery or serfdom, discipline allows for the increase of the economic productivity of the body, while at the same time weakening its forces to assure political subjection. It is exactly this coupling of economic and political imperatives that define discipline and establish its status as a technology:

> The historical moment of the disciplines was the moment when an art of the human body was born, which was directed not only at the growth of its skills, nor at the intensification of its subjection, but at the formation of a relation that in the mechanism itself makes it more obedient as it becomes more useful, and conversely. (Foucault 1977, 137–138)

In the second half of the 18th century another technology of power emerged, which was directed not at the bodies of individuals but at the collective body of a population. By "population" Foucault does not imagine a legal or political entity (e.g., the totality of individuals) but an independent biological corpus: a "social body" that is characterized by its own processes and phenomena, such as birth and death rates, health status, life span, and the production of wealth and its circulation. The totality of the concrete processes of life in a population is the target of a "technology of security" (2003, 249). This technology aims at the mass phenomena characteristic of a population and its conditions of variation in order to prevent or compensate for dangers and risks that result from the existence of a population as a biological entity. The instruments applied here are regulation and control, rather than discipline and supervision. They define a "technology which aims to establish a sort of homeostasis, not by training individuals but by achieving an overall equilibrium that protects the security of the whole from internal dangers" (ibid., 249).

Disciplinary technology and security technology differ not only in their objectives and instruments and the date of their historical appearance but also in where they are situated institutionally. Disciplines developed inside of institutions, such as the army, prisons, schools, and hospitals, whereas the state organized and centralized the regulation of the population from the 18th century on. The collection of demographic data was important in this regard, as were the tabulation of resources and statistical censuses related to life expectancy and the frequency of illness. Two series, therefore, may be discerned: "the body–organism–discipline–institution series, and the population–biological processes–regulatory mechanisms–State" (ibid., 250).

The difference between the two components of biopolitics should, however, be acknowledged with caution. Foucault stresses that discipline and control form "two poles of development linked together by a whole intermediary cluster of relations" (1980, 139). They are not independent entities but define each other. Accordingly, discipline is not a form of individualization that is applied to already existing individuals, but rather it presupposes a multiplicity.

Similarly, population constitutes the combination and aggregation of individualized patterns of existence to a new political form. It follows that "individual" and "mass" are not extremes but rather two sides of a global political technology that simultaneously aims at the control of the human as individual body and at the human as species (see Foucault 2003, 242–243). Moreover, the distinction between the two political technologies cannot be maintained for historical reasons. For example, the police in the 18th century operated as a disciplinary apparatus and as a state apparatus. State regulation in the 19th century relied on a range of institutions in civic society, such as insurance, medical-hygienic institutions, mutual aid associations, philanthropic societies, and so on. In the course of the 19th century it is possible to observe alliances between the two types of power that Foucault describes as "apparatuses" (*dispositifs*).

According to Foucault, the "apparatus of sexuality"—whose investigation stands at the center of *The History of Sexuality, Vol. 1*— occupies a prominent position in this setting. Foucault is interested in sexuality because of its position "at the pivot of the two axes" between both forms of power (1980, 145). Sexuality represents a bodily behavior that gives rise to normative expectations and is open to measures of surveillance and discipline. At the same time, it is also important for reproductive purposes and as such part of the biological processes of a population (cf. Foucault 2003, 251–252). Thus, sexuality assumes a privileged position since its effects are situated on the microlevel of the body and on the macrolevel of a population. On the one hand, it is taken to be the "stamp of individuality": "behind" the visible behavior, "underneath" the words spoken, and "in" the dreams one seeks hidden desires and sexual motives. On the other hand, sexuality has become "the theme of political operations, economic interventions . . . , and ideological campaigns for raising standards of morality and responsibility: it was put forward as the index of a society's strength, revealing of both its political energy and its biological vigor" (1980, 146).

In this context, the concept of the norm plays a key role. The ancient "power over life and death" operated on the basis of the binary legal code, whereas biopolitics marks a movement in which the "right" is more and more displaced by the "norm." The absolute right of the sovereign tends to be replaced by a relative logic of calculating, measuring, and comparing. A society defined by natural law is superseded by a "normalizing society":

> It is no longer a matter of bringing death into play in the field of sovereignty, but of distributing the living in the domain of value and utility. Such a power has to qualify, measure, appraise, and hierarchize, rather than display itself in its murderous splendor; it does not have to draw the line that separates the enemy of the sovereign from his loyal subjects. It effects distributions around the norm. (1980, 144)

However, Foucault's thesis that modern politics tends to become biopolitics does not imply that sovereignty and the "power over death" play no role any more. On the contrary, the sovereign "right of death" has not disappeared but is subordinated to a power that seeks to maintain, develop, and manage life. As a consequence, the power over death is freed from all existing boundaries, since it is supposed to serve the interest of life. What is at stake is no longer the juridical existence of a sovereign but rather the biological survival of a population. The paradox of biopolitics is that to the same degree to which the security and the amelioration of life became an issue for political authorities, life is threatened by hitherto unimaginable technical and political means of destruction:

> Wars were never as bloody as they have been since the nineteenth century, and . . . never before did the regimes visit such holocausts on their own populations. . . . Entire populations are mobilized for the purpose of wholesale slaughter in the name of life necessity: massacres have become vital. It is as managers of life and survival, of bodies and the race, that so many regimes have been able to wage so many wars, causing so many men to be killed. (1980, 136–137)

Foucault sees the reason for this in modern racism, which ensures the "death-function in the economy of biopower" (2003, 258).

Racism and Power of Death

Whereas the difference between sovereign power and biopower is central to *The History of Sexuality, Vol. 1*, Foucault chooses another starting point in his 1976 lectures at the Collège de France. Biopolitics here stands not so much for the "biological threshold of modernity" (1980, 143) as for the "break between what must live and what must die" (2003, 254). Foucault's working thesis is that the transformation of sovereign power into biopower leads to a shift from a political-military discourse into a racist-biological one. The political-military discourse was present in the 17th and 18th centuries. It strove to be a "challenge to royal power" (ibid., 58), emerging in the Puritan rebellion of prerevolutionary England and a bit later in France with the aristocratic opposition to King Louis XIV. Very early in this process the expression "race" emerged, which was not yet linked to a biological signification. Rather, it initially described a specific historical-political division. Fundamental was the idea that society is divided into two hostile camps and two antagonistic social groups that coexist on a territory without mixing and that clearly distinguish themselves from one another through, for example, geographical origin, language, or religion. This "counterdiscourse" principally contested the legitimacy of sovereign power and the postulated universality of laws, which it unmasked as the specific norms and forms of tyranny.

In the 19th century, according to Foucault, this historical-critical discourse experienced "two transcriptions" (ibid., 60). The discourse of "race war" experienced first an "openly biological transcription" that, even before Darwin, drew on elements of materialist anatomy and physiology (ibid.). This historical-biological race theory conceives of societal conflicts as "struggles for existence" and analyzes them in the light of an evolutionary schema. In a second transformation, "race war" is interpreted as class struggle and investigated according to the principle of dialectics. At the beginning of the 19th century, a revolutionary discourse emerged in which the problem of politically determined "race" was increasingly replaced by the thematic of social class (ibid., 61, 78–80).

Foucault argues that the two "reformulations" of the political problematic of the "race war" at the end of the 19th century result in a biological-social discourse. This "racism" (only in the 19th century does this term acquire its current meaning) draws on elements of the biological version in order to formulate an answer to the social revolutionary challenge. In place of the historical-political thematic of war, with its slaughters, victories, and defeats, enters the evolutionary-biological model of the struggle for life. According to Foucault, this "dynamic racism" (1980, 125) is of "vital importance" (2003, 256) because it furnishes a technology that secures the function of killing under the conditions of biopower: "How can a power such as this kill, if it is true that its basic function is to improve life, to prolong its duration, to improve its chances, to avoid accidents, and to compensate for failings? . . . It is . . . at this point that racism intervenes" (ibid., 254).

Racism fulfills two important functions within an economy of biopower. First, it creates fissures in the social domain that allow for the division of what is imagined in principle to be a homogeneous biological whole (for example, a population or the entire human species). In this manner, a differentiation into good and bad, higher and lower, ascending or descending "races" is made possible and a dividing line established "between what must live and what must die" (ibid., 254).[1] Indeed, "to fragment, to create caesuras within the biological continuum" presupposes its creation (ibid. 255). In contrast to the traditional theme of race war, which is marked by the idea of a binary society divided into two opposing races, in the 19th century there emerged the idea of a society "that is, in contrast, biologically monist" (ibid., 80). The idea of a plurality of races shifts to one of a single race that is no longer threatened from without but from within. The result is a "racism that society will direct against itself, against its own elements, and its own products. This is the internal racism of permanent purification, and it will become one of the basic dimensions of social normalization" (ibid., 62). From this perspective, homogenization and hierarchization do not oppose one another but rather represent complementary strategies.

The second function of racism goes even further. It does not limit itself to establishing a dividing line between "healthy" and "sick," "worthy of living" and "not worthy of living." Rather, it searches for "the establishment of a positive relation of this type: 'The more you kill, the more deaths you will cause' or 'The very fact that you let more die will allow you to live more'" (ibid., 255). Racism facilitates, therefore, a dynamic relation between the life of one person and the death of another. It not only allows for a hierarchization of "those who are worthy of living" but also situates the health of one person in a direct relationship with the disappearance of another. It furnishes the ideological foundation for identifying, excluding, combating, and even murdering others, all in the name of improving life: "The fact that the other dies does not mean simply that I live in the sense that his death guarantees my safety; the death of the other, the death of the bad race, of the inferior race (or the degenerate, or the abnormal) is something that will make life in general healthier" (ibid., 255).

The idea of society as a biological whole assumes the provision of a central authority that governs and controls it, watches over its purity, and is strong enough to confront "enemies" within its borders and beyond: the modern state. Foucault argues that, from the end of the 19th century, at the latest, racism guided the rationality of state actions; it finds form in its political instruments and concrete policies as "State racism" (ibid., 261). While the historico-political discourse of race was still directed against the state and its apparatuses (which it denounces as the instruments of domination of one group over another) and against its laws (whose partisanship it unmasks), then the discourse of race ultimately places a weapon in the hands of the state:

the State is no longer an instrument that one race uses against another: the state is, and must be, the protector of the integrity, the superiority, and the purity of the race. The idea of racial purity, with all its monistic, Statist, and biological implications: that is what replaces the idea of race struggle. I think that racism is born at the point when the theme of racial purity replaces that of race struggle. (2003, 81)

Foucault points out two further transformations of racist discourse in the 20th century: Nazi Germany and the state socialism of the Soviet Union. National Socialism harked back to motifs of the old race war in order to launch imperialist expansion outward and to attack its internal enemies. It is characterized by an "oneiric exaltation of a superior blood [that] implied both the systematic genocide of others and the risk of exposing oneself to a total sacrifice" (1980, 150). Soviet racism, however, lacked this theatrical moment. It instead deployed the discrete means of a medical police force. The utopia of a classless society was to be realized in state socialism through the project of cleansing a society in which all those who diverged from the dominant ideology were treated as either "sick" or "crazy." In this variant of state racism, class enemies became biologically dangerous and had to be removed from the social body (2003, 82–83).

Foucault's analysis of racism has been rightly criticized as being limited and selective. Although the problem of colonialism is mentioned cursorily in his discussion, it is not handled in a systematic manner. Foucault neither recognizes the inner interrelationship of nation, citizenship, and racism, nor is he interested in the sexual component of the race discourse.[2] Despite these lacunae and deficits, it is clear that Foucault's genealogy of modern racism contains a range of analytical assets. First, he conceives of racism neither as an ideological construct nor as an exceptional situation nor as a response to social crises. According to Foucault, racism is an expression of a schism within society that is provoked by the biopolitical idea of an ongoing

and always incomplete cleansing of the social body. Racism is not defined by individual action. Rather, it structures social fields of action,
guides political practices, and is realized through state apparatuses.

Furthermore, Foucault challenges the traditional political demarcation between conservative and critical positions. The old notion of
race war was a discourse that directed itself against established sovereign power and its self-representation and principles of legitimation.
Through the "transcriptions" Foucault identifies (ibid., 60), the political project of liberation turns into one of racist concern with biological purity; the prophetic-revolutionary promise becomes medical-hygienic conformity with the norm; from the struggle against
society and its constraints, there follows the imperative to "defend
society" against biological dangers; a discourse against power is
transformed into a discourse of power: "Racism is, quite literally,
revolutionary discourse in an inverted form" (ibid., 81). Foucault's
analysis draws attention to "tactical polyvalence" (1980, 100) and the
inner capacity for transformation that race discourse contains. In this
way it becomes possible to account for some contemporary neoracist
strategies that do not so much stress biological difference but rather
assert the allegedly fundamental cultural differences between ethnic
groups, peoples, or social groups.

Political Economy and Liberal Government

Foucault's 1978 and 1979 lectures at the Collège de France place the
theme of biopolitics in a more complex theoretical framework. In the
course of the lectures he examines the "genesis of a political knowledge" of guiding humans beings from antiquity via the early modern notion of state reason and "police science" (*Polizeywissenschaft*)
to liberal and neoliberal theories (2007, 363). Central to these is the
concept of government. Foucault proposes a "very broad meaning"
of the term, taking up the diversity of meanings that it carried well
into the 18th century (2000b, 341). Although the word has a purely
political meaning today, Foucault shows that up until well into the

18th century the problem of government was placed in a more general context. Government was a term discussed not only in political tracts but also in philosophical, religious, medical, and pedagogic texts. In addition to management by the state or administration, government also addressed problems of self-control, guidance for the family and for children, management of the household, directing the soul, and other questions.[3]

Within this analytics of government, biopolitics takes on a decisive meaning. The "birth of biopolitics" (the title of the 1979 lecture series) is closely linked to the emergence of liberal forms of government. Foucault conceives of liberalism not as an economic theory or a political ideology but as a specific art of governing human beings. Liberalism introduces a rationality of government that differs both from medieval concepts of domination and from early modern state reason: the idea of a nature of society that constitutes the basis and the border of governmental practice.

This concept of nature is not a carryover of tradition or a premodern relic but rather a marker of a significant historical rupture in the history of political thought. In the Middle Ages, a good government was part of a natural order willed by God. State reason breaks with this idea of nature, which limited political action and embedded it in a cosmological continuum. Instead, state reason proposes the artificiality of a "leviathan"—which provokes the charge of atheism. With the Physiocrats and political economy, nature reappears as a point of reference for political action. However, this is a different nature that has nothing to do with a divine order of creation or cosmological principles. At the center of liberal reflection is a hitherto unknown nature, the historical result of radically transformed relations of living and production: the "second nature" of the evolving civil society (see Foucault 2007).

Political economy, which emerged as a distinctive form of knowledge in the 18th century, replaced the moralistic and rigid principles of mercantilist and cameralist economic regulation with the idea of spontaneous self-regulation of the market on the basis of "natural" prices. Authors such as Adam Smith, David Hume, and Adam Ferguson assumed that there exists a nature that is peculiar to governmental practices and that governments have to respect this nature in their operations. Thus, governmental practices should be in line with the laws of a nature that they themselves have constituted. For this reason, the principle of government shifts from external congruence to internal regulation. The coordinates of governmental action are no longer legitimacy or illegitimacy but success or failure; reflection focuses not on the abuse or arrogance of power but rather on ignorance concerning its use.

Thus, for the first time political economy introduces into the art of government the question of truth and the principle of self-limitation. As a consequence, it is no longer important to know whether the prince governs according to divine, natural, or moral laws; rather, it is necessary to investigate the "natural order of things" that defines both the foundations and the limits of governmental action. The new art of government, which became apparent in the middle of the 18th century, no longer seeks to maximize the powers of the state. Instead, it operates through an "economic government" that analyzes governmental action to find out whether it is necessary and useful or superfluous or even harmful. The liberal art of government takes society rather than true state as its starting point and asks, "Why must one govern? That is to say: What makes government necessary, and what ends must it pursue with regard to society in order to justify its own existence?" (2008, 319).

A reduction of state power in no way follows from this historical shift, however. Paradoxically, the liberal recourse to nature makes it possible to leave nature behind or, more precisely, to leave behind a certain concept of nature that conceives of it as eternal, holy, or unchangeable. For liberals, nature is not an autonomous domain in which intervention is impossible or forbidden as a matter of principle. Nature is not a material substratum to which governmental practices are applied but rather their permanent correlate. It is true that there is a "natural" limit to state intervention, as it has to take into account the nature of the social facts. However, this dividing line is not a negative borderline, since it is precisely the "nature" of the population that opens up a series of hitherto unknown possibilities of intervention. These do not necessarily take the form of direct interdictions or regulations: "laisser-faire," inciting, and stimulating become more important than dominating, prescribing, and decreeing (2007, 70–76; 2008, 267–316).

In this context, Foucault gives a new meaning to the concept of technologies of security, which he used in earlier works. He regards security mechanisms as counterparts to liberal freedom and as the condition for its existence. Security mechanisms are meant to secure and protect the permanently endangered naturalness of the population, as well as its own forms of free and spontaneous self-regulation. Foucault distinguishes analytically between legal regulations, disciplinary mechanisms, and technologies of security. Legal normativity operates by laws that codify norms, whereas discipline installs hierarchical differentiations that establish a division between those considered normal and abnormal, suitable and capable, and the others. It functions by designing an optimal model and its operationalization, that is, by employing techniques and procedures to adjust and adapt individuals to this standard.

The technologies of security represent the very opposite of the disciplinary system: whereas the latter assumes a prescriptive norm, the former take the empirical norm as a starting point, which serves as a regulative norm and allows for further differentiations and variations. Rather than adjusting reality to a predefined "should-be" value, the technologies of security take reality as the norm: as a statistical distribution of events, as average rate of diseases, births and deaths, and so on. They do not draw an absolute borderline between the permitted and the prohibited; rather, they specify an optimal middle within a spectrum of variations (2007, 55–63).

The formation of political economy and population as new political figures in the 18th century cannot be separated from the emergence of modern biology. Liberal concepts of autonomy and freedom are closely connected to biological notions of self-regulation and self-preservation that prevailed against the hitherto dominant physical-mechanistic paradigm of investigating bodies. Biology, which emerged about 1800 as the science of life, assumes a basic principle of organization that accounts for the contingency of life without any foundational or fixed program. The idea of an external order that corresponds to the plans of a higher authority beyond life is displaced by the concept of an inner organization, whereby life functions as a dynamic and abstract principle common to all organisms. From this point on, such categories as self-preservation, reproduction, and development (cf. Foucault 1970) serve to characterize the nature of living bodies, which now more clearly than ever before are distinguishable from artificial entities.

In the 1978 and 1979 lectures, Foucault conceives of "liberalism as the general framework of biopolitics" (2008, 22). This account of liberalism signals a shift of emphasis in relation to his previous work. The theoretical displacement results from the self-critical insight that his earlier analysis of biopolitics was one-dimensional and reductive, in the sense that it primarily focused on the biological and physical life of a population and on the politics of the body. Introducing the notion of government helps to broaden the theoretical horizon, as it links the interest in a "political anatomy of the human body" with the investigation of subjectivation processes and moral-political forms of existence. From this perspective, biopolitics represents a particular and dynamic constellation that characterizes liberal government. With liberalism, but not before, the question arises of how subjects are to be governed if they are both legal persons and living beings (see ibid. 2008, 317). Foucault focuses on this problem when he insists that biopolitical problems cannot be separated from the framework of political rationality within which they appeared and took on their intensity. This means "liberalism," since it was in relation to liberalism that they assumed the form of a challenge. How can the phenomena of "population," with its specific effects and problems, be taken into account in a system concerned about respect for legal subjects and individual free enterprise? In the name of what and according to what rules can it be managed? (2008, 317)

The reformulation of the concept of biopolitics within an analytics of government has a number of theoretical advantages. First, such a research perspective allows for the exploration of the connections between physical being and moral-political existence: how do certain objects of knowledge and experiences become a moral, political, or legal problem? This is the theme of the last volume of Foucault's *History of Sexuality*, at whose center stand moral problematizations of physical experiences and forms of self-constitution (1988, 1990). Contemporary examples are the figure of the human being and the legal construct of human dignity, both of which are coming under increasing pressure as a result of biotechnical innovation. The problem has thus emerged, for example, of whether embryos possess human dignity and can claim human rights. Furthermore, on what "natural" assumptions do the guarantees of political and social rights depend? What is the relationship between different forms of socialization and biological traits? Such a perspective focuses our attention on the relationship between technologies and governmental practices: How do liberal forms of government make use of corporeal techniques and forms of self-guidance? How do they form interests, needs, and structures of preference? How do present technologies model individuals as active and free citizens, as members of self-managing communities and organizations, as autonomous actors who are in the position—or at least should be—to rationally calculate their own life risks? In neoliberal theories, what is the relationship between the concept of the responsible and rational subject and that of human life as human capital?

Foucault's writing did not so much systematically pursue this analytic perspective as offer promising suggestions for its development. He never made his remarks on the relation between biopolitics and liberalism concrete—a project that was meant to stand at the center of the 1979 lecture (see 2008, 21–22, 78). Regrettably, what we are left with is the "intention," as Foucault conceded self-critically in the course of the lecture (ibid., 185–186).

Resistance and the Practices of Freedom

Foucault's interest in liberal government also leads him to a modified appraisal of resistance and practices of freedom that he now conceives of as an "organic" element of biopolitical strategies. According to him, processes of power that seek to regulate and control life provoke forms of opposition, which formulate claims and demand recognition in the name of the body and of life. The expansion and intensification of control over life makes it at the same time the target of social struggles:

> [A]gainst this power . . . the forces that resisted relied for support on the very thing it invested, that is, on life and man as a living being. . . . [W]hat was demanded and what served as an objective was life, understood as the basic needs, man's concrete essence, the realization of his potential, a plenitude of the possible. Whether it was Utopia that was wanted is of little importance; what we have seen has been a very real process of struggle; life as a political struggle was in a sense taken at face value and turned back against the system that was bent on controlling it. (1980, 144–145)

The disciplining of bodies and the regulation of the population caused new political struggles that did not invoke old and forgotten rights but claimed new categories of rights, such as the right to life, a body, health, sexuality, and the satisfaction of basic needs. Foucault's historical thesis is that biopolitical conflicts have become increasingly important since World War II and especially since the 1960s. Alongside the struggles against political, social, or religious forms of domination and economic exploitation, a new field of conflicts emerged: struggles against forms of subjectivation (see 2000b, 331–332). It is possible to detect a "developing crisis of government" (2000c, 295), which manifests itself in numerous social oppositions between men and women, conflicts on the definition of health and disease, reason and madness, in the rise of ecological movements, peace movements, and sexual minorities. Taken together these developments signal that traditional forms of subjectivation and concepts of the body are losing their binding force. These struggles are characterized by the fact that they oppose a "government of individualization" (2000b, 330). They call into question the adaptation of individuals to allegedly universally valid and scientifically grounded social norms that regulate models of the body, relations of the sexes, and forms of life.

In Foucault's last works, he analyzes ancient self-practices in the context of his book project on the "history of sexuality." Even if the notion of biopolitics no longer occupies a strategic role in his writings of that time, he continues to be interested in forms of resistance against a governmental technology that has human life as its object. Against this "naturalization" of power, with its reference to the apparently self-evident and universal normative claims of biological life, Foucault proposed to understand human life rather as a "work of art." With his analysis of the ancient "aesthetics of existence," he sought to reactivate a new "art of living" that could move beyond the truth claims of both the life sciences and the human sciences (cf. Foucault 1988, 1990).

Foucault's concept of biopolitics was, after his death in 1984, received in many different ways. Two diametrically opposed interpretations have become increasingly influential in recent years. Both draw attention to lacunae in and problems with Foucault's framing of biopolitics and aim to develop the concept further. However, the diagnoses of the problems are as diverse as the suggested solutions. On the one hand are the writings of Giorgio Agamben, and on the other are the works of Michael Hardt and Antonio Negri, both of which will be introduced in the following chapters.

4

Sovereign Power and Bare Life: Giorgio Agamben

FOR SOME TIME now, the work of Italian philosopher Giorgio Agamben has been receiving growing attention and appreciation.[1] Yet it was only with the appearance of *Homo Sacer* in 1995 that he became known to a wider audience (Agamben 1998). The book was an international bestseller, and its author became an intellectual star. The reason for this lay not least in the work's brilliance in bringing together philosophical reflection with political critique. Above all, however, his fundamental thesis is provocative enough to have earned him greater notice outside of philosophical circles. For Agamben asserts nothing less than the "inner solidarity between democracy and totalitarianism" (ibid., 10) and defines the concentration camp as the "biopolitical paradigm of the West" (ibid., 181).

Homo Sacer is the first volume of a four-volume work of which further volumes have in the meantime appeared and in which Agamben expands and concretizes his thesis. In these works, Agamben reads the present as the catastrophic terminus of a political tradition that has its origins in ancient Greece and that led to the Nazi concentration camps. Whereas the advent of biopolitical mechanisms in the 17th and 18th centuries signaled for Foucault a historical caesura, Agamben insists on a logical connection between sovereign power and biopolitics. That is, biopolitics forms the core of the sovereign practice of power. The modern era signifies, accordingly, not a break with the Western tradition but rather a generalization and radicalization of that which was simply there at the beginning. According to

Agamben, the constitution of sovereign power assumes the creation of a biopolitical body. Inclusion in political society is only possible, he writes, through the simultaneous exclusion of human beings who are denied full legal status.

In what follows, I present Agamben's revision of Foucault's conception of biopolitics and discuss its analytical merits as well as its limits. The first section briefly presents Agamben's initial thesis, and the second part investigates its diagnostic potential for an analysis of contemporary societies. In the third section, I identify several theoretical problems posed by Agamben's conception of biopolitics, including his implicit adherence to a juridical conception of power, his fixation on the state, his neglect of socioeconomic aspects of the biopolitical problematic, and the quasi-ontological foundation of his theoretical model.

The Rule of the Exception

Agamben takes up not only Foucault's works but also those of Carl Schmitt, Walter Benjamin, Hannah Arendt, Martin Heidegger, and Georges Bataille. He begins with a distinction that he believes has determined the occidental political tradition since Greek antiquity. The central binary relationship of the political is not that between friend and enemy but rather the separation of bare life (*zoé*) and political existence (*bíos*)—that is, the distinction between natural being and the legal existence of a person. According to Agamben, we find at the beginning of all politics the establishment of a borderline and the inauguration of a space that is deprived of the protection of the law: "The original juridico-political relationship is the ban" (1998, 181).

Agamben outlines this hidden foundation of sovereignty through a figure he derives from archaic Roman law: *homo sacer*. This is a person whom one could kill with impunity, since he was banned from the politico-legal community and reduced to the status of his physical existence. For Agamben, this obscure figure represents the other side of the logic of sovereignty. "Bare life," which is considered to be marginal and seems to be furthest from the political, proves to be the solid basis of a political body, which makes the life and death of a human being the object of a sovereign decision. From this perspective, the production of *homines sacri* represents a renounced yet constitutive part of Western political history.

The trace of *homo sacer* runs from Roman exiles through the condemned of the Middle Ages to the inmates of Nazi camps, and beyond. In contemporary times, Agamben conceives of "bare life" as existing, for example, in asylum seekers, refugees, and the brain dead. These apparently unrelated "cases" have one thing in common: although they all involve human life, they are excluded from the protection of the law. They remain either turned over to humanitarian assistance and unable to assert a legal claim or are reduced to the status of "biomass" through the authority of scientific interpretations and definitions.[2]

Agamben's reconstruction of the interrelationships between sovereign rule and biopolitical exception results in an unsettling outcome. The thesis of the concentration camp as "the hidden matrix of the politics in which we still live" (Agamben 2000, 44) makes claims for an inner link between the emergence of human rights and the development of concentration camps. In this sense, there is no sharp division between parliamentary democracies and totalitarian dictatorships, liberal constitutional states and authoritarian regimes. Agamben's claim of an "inner solidarity between democracy and totalitarianism" (1998, 10) has provoked much resistance. Although his thesis of the camp as "biopolitical paradigm of the modern" (ibid., 117) in no way makes relative or trivializes Nazi extermination policies, it remains the case that Agamben ignores important and essential differences. The criticism that Agamben "levels" differences is a less relevant argument than his lack of concretization and the excessive dramatization that may lead, ultimately, to the impression that *homo sacer* is "forever and everywhere" (Werber 2002, 622).

"Bare Life" and the Camp

What does Agamben mean when he describes the concentration camp as the "hidden paradigm of the political space of modernity" (1998, 123)? Evidently, the camp for him does not so much represent a concrete historical place or a defined spatial unity, but symbolizes and fixes the border between "bare life" and political existence. The camps in this sense are not only Nazi concentration camps or contemporary deportation centers but rather any space in which "bare life" is systematically produced: *the camp is the space that is opened when the state of exception begins to become the rule* (ibid., 168–169, emphasis in original). Agamben sees in the camps the "hidden matrix" (ibid., 175) of the political domain, and he wants to make visible the underlying logic in order to better conceive the present political constellation. In other words, Agamben proposes a significantly new definition of the "camp," one that displaces the traditional definition. The camp, once the epitome and manifestation of the difference between friend and enemy, is turned by Agamben into the "materialization of the state of exception" (ibid., 174), where law and factum, rule and exception, indistinguishably commingle.

In contrast to Foucault, Agamben proceeds from a fundamental continuity of biopolitical mechanisms whose foundation he finds in the logic of sovereignty. Yet he also uncovers a historical caesura. The modern era, he writes, distinguishes itself from previous ones to the extent that "bare life," formerly on the margins of political existence, now increasingly shifts into the center of the political domain. The threshold to biopolitical modernity will be crossed, according to Agamben, when bare life proceeds beyond the state of exception to become central to political strategies; the exception will become the rule, and the difference between inside and outside, factum and law enter into a "zone of irreducible indistinction" (ibid., 9). Modern biopolitics, writes Agamben, has "two faces":

> [T]he spaces, the liberties, and the rights won by individuals in their conflicts with central powers always simultaneously prepared a tacit but increasing inscription of individuals' lives within the state order, thus offering a new and more dreadful foundation for the very sovereign power from which they wanted to liberate themselves. (Ibid., 121)

It is this same "bare life" that in democracies results in the private having priority over the public and that in totalitarian states becomes a decisive political criterion for the suspension of individual rights.

But even if the same substratum ("bare life") forms the foundation of each form of government, this does not mean that they should all be assessed as politically the same. In contrast to what most commentators argue, Agamben is in no way equating democracy and dictatorship or devaluing civil freedoms or social rights. Rather, he argues that democratic rule of law is not an alternative political project to Nazi or Stalinist dictatorships. These political regimes, rather, radicalize biopolitical trends that according to Agamben are already found in other political contexts and historical epochs and whose power today has increased rather than decreased.

Thus, Agamben does not follow a logic of oversimplified parallels. Rather, he tries to elucidate the common ground for these very different forms of government, namely, the production of "bare life." Instead of insisting that the Nazi camps represent a logical exception or a historically marginal phenomenon, he searches instead for the "regularity" or normality of this exception and asks to what extent "bare life" is an essential component of contemporary political rationality, since life and its preservation and prolongation are increasingly the object of legal regulations (2000, 37–45).

Agamben sees an intensification of the biopolitical problematic after the end of both the Nazi and Stalinist dictatorships. He argues that since then biopolitics has "passed beyond a new threshold": "in modern democracies it is possible to state in public what the Nazi biopoliticians did not dare to say" (1998, 165). Whereas Nazi biopolitics still concentrated on identifiable individuals or definite subpopulations, Agamben argues that "in our age all citizens can be said, in a specific but extremely real sense, to appear virtually as *homines sacri*" (ibid., 111,). Evidently, Agamben assumes that the boundary that once ran between individuals or social groups is now incorporated in individual bodies and is, to a certain extent, internalized. The boundary between politically relevant existence and bare life has today moved necessarily "inside every human life. . . . Bare life is no longer confined to a particular place or a definite category. It now dwells in the biological body of every living being" (ibid., 140).

Unfortunately, Agamben leaves this aggravation of the biopolitical problem extremely vague. In place of conceptual work and historical sensibility, one frequently finds a hunt for aporias and a tendency toward subsumption. His thesis that rule and exception "enter a zone of absolute indeterminacy" (2000, 42) is coupled with a lack of conceptual differentiation. Even if all subjects are *homines sacri*, they are so in very different ways. Agamben detracts from his argument by stating that·everyone is susceptible to being reduced to the status of "bare life"—without clarifying the mechanism of differentiation that distinguishes between different values of life. It remains very unclear to what extent and in what manner comatose patients in hospital share the fate of prisoners in concentration camps, whether the asylum seekers in prisons are bare life to the same degree and in the same sense as the Jews in the Nazi camps. If on the one hand Agamben seems to tend toward an exaggerated dramatization rather than a sober assessment—he even regards people killed on motorways indirectly as *homines sacri* (1998, 114; Werber 2002, 422)—then on the other hand his analysis must tolerate criticism that it represents an unacceptable trivialization—that Auschwitz serves him as an object lesson that perpetually renews itself (cf. Agamben 1999, 133–134, 156).

Three Problems

This meager capacity for differentiation is not a coincidental flaw in argumentation but rather the inevitable result of an analysis that deals with biopolitics from a one-sided and abbreviated perspective. Three sets of problems are particularly apparent in Agamben's work: the juridical, the state-centric, and the quasi-ontological framing of biopolitics.

Where the first juridical charge of a fixation on legal questions is concerned, one notices that Agamben conceives of the "camp" not as a differentiated and differentiating continuum but simply as a "line" (1998, 122) that more or less unambiguously divides bare life and political existence. His attention is directed solely toward the establishment of a border—a border that he comprehends not as a tiered or graded zone but as a line without extension or dimension that reduces the question to an either-or. Within these parameters, he can no longer analyze how gradations and valuations within "bare life" emerge, how life can be qualified as "higher" or "lower," as "descending" or "ascending." These processes of differentiation evade him, for he is interested not so much in "life" as in its "bareness." Discipline and training, the normalization and standardization of life, are not central to his thinking. Instead, death as the establishment and materialization of a boundary is. For Agamben, biopolitics is therefore above all "thanatopolitics" (1998, 142; cf. 1999, 84–86; Fitzpatrick 2001, 263–265; Werber 2002, 419).[3]

This point represents the crux of the difference between Agamben and Foucault. Foucault shows that sovereign power is by no means sovereign, since its legitimacy and efficiency depend on a "microphysics of power," whereas in Agamben's work sovereignty produces and dominates bare life. For Agamben "*the production of a biopolitical body is the original activity of sovereign power*" (1998, 6; emphasis in original). The binary juxtaposition of *bíos* and *zoé*, political existence and bare life, rule and exception, refers to that juridical model of power that Foucault criticizes. Agamben's analysis remains in thrall to the law and owes more to Carl Schmitt than it does to Foucault. If Schmitt situates sovereign power in its ability to call for a state of exception and a suspension of rights (cf. Schmitt 1996), Foucault is interested in normal conditions, which exist below, next to, and partially counter to legal mechanisms. Whereas Schmitt is interested in how the norm is suspended, Foucault is concerned with the production of normality (Foucault 2003, 23–41; Fitzpatrick 2001, 259–261; Deuber-Mankowsky 2002, 108–114).

In focusing on law and the sovereign right of banishment, Agamben banishes central aspects of biopolitics from his analysis. He suggests that the state of exception is not only the origin of politics but also its very purpose and definition. With this configuration, politics would exhaust itself in the production of *homines sacri,* which must be regarded as unproductive, for "bare life" is created only to be oppressed and killed. Agamben dismisses the fact that biopolitical interventions in no way limit themselves to the processing of the opposition between biological and political existence. Instead of simply exterminating "bare life" or allowing it to be killed with impunity, these interventions subordinate it to a "bioeconomic" imperative of increasing value whose aim is to improve chances of survival and the quality of life. In other words, Agamben fails to recognize that biopolitics is essentially a political economy of life. His analysis remains under the spell of sovereign power and blind to all mechanisms that operate outside the law.

Contrary to what Agamben assumes, biopolitical mechanisms do not concentrate on that which is reduced to the status of a living entity and whose elemental rights are withheld. The analysis of biopolitics cannot be limited to those without legal rights, such as the refugee or the asylum seeker, but must encompass all those who are confronted with social processes of exclusion (even if they may formally enjoy full political rights), namely, the "useless," the "unnecessary," or the "redundant." Whereas in the past these figures inhabited only peripheral spaces, today in a global economy these forms of exclusion can also be found in the industrialized centers in which social questions are newly posed because of the dismantling of the welfare state and the crisis of the labor economy.

The second problem with Agamben's analysis consists in its concentration on state apparatuses and centralized forms of regulation. His thesis concerning the further politicization of nature is plausible, since today the beginning, continuation, and end of life processes are, through biotechnological and medical innovations, susceptible to decision-making processes. However, his focus on Nazi race politics leads to a distorted view of the present. Agamben is apparently not aware that biopolitics is not only the purview of government regulation. It is also a field of "autonomous" subjects who as rational patients, entrepreneurial individuals, and responsible parents (should) demand biotechnological options. Less and less frequently does the state, due to its concern with the health of the "people's body" (*Volkskörper*), decide who is worthy of living. Increasingly these decisions are handed over to individuals. The determination of "quality of life" has become a question of individual utility, personal preferences, and the suitable allocation of resources.

The principle danger today is not that the body or its organs will succumb to state control (cf. Agamben 1998, 164–165). On the contrary, the danger is that the state will, in the name of "deregulation," retreat from the domains it once occupied in society and hand over decisions pertaining to the value of life and determinations of when it begins and ends to the realm of science and commercial interests, as well as to the deliberations of ethics committees, expert commissions, and citizen panels.

This "withdrawal of the state" could itself be analyzed as a political strategy, albeit one that does not necessarily refuse individuals legal rights. While the suspension of legal rights might remain important in determining who is allowed to become part of a community and who is eligible for legal rights at all, the political strategy that shifts legal and regulatory competencies from the public and legal domain to the private sphere will probably pose a much greater threat in the future. This tendency not only manifests itself in the current possibility of privately appropriating such corporeal substances as genes or cell lines and using them commercially, but also is hinted at by the examples Agamben chooses, such as assisted suicide and transplantation medicine. It is to be expected that in the future living wills and contractual arrangements will take the place of explicit state prescriptions or proscriptions.

Finally, a third point needs to be considered. Agamben uses a quasi-ontological concept of biopolitics, so that his notion of life remains curiously static and ahistorical. Indeed, Agamben notes that "bare life" does not refer to a natural or presocial condition, though it seems to emerge from a kind of life "substance" that is historically modeled and modified when he writes of "bare life as such" (1998, 4). The notion of a continuity between a biopolitics situated in antiquity and the present is unconvincing. The term "life" as it is used in antiquity and modernity has little but a name in common, and this is so because "life" is a specifically modern concept. Until the second half of the 18th century, the strict differentiation between a natural being and an artificial production, between organic and inorganic, was unknown. Only with the appearance of modern biology was "life" or the "life force" granted an identity as an independent working principle that described the emergence, preservation, and development of natural bodies—a principle distinguished by its own autonomous laws and its own area of study. Until the 18th century, philosophy and science assumed a continuity between the natural and the artificial. Beginning in the 1700s, there was a strict division between the two. Whereas the artificial was traced back to an agent of causality and was deemed to be governed from the outside, this did not apply to the inner teleology of living entities. Life was from the 18th century on conceived of as a form of self-organization that obeyed only "inner causes."

Agamben's attempt to correct and amend Foucault (cf. 1998, 9) also abandons the latter's central insight, namely, that biopolitics is a historical phenomenon that cannot be separated from the development of modern states, the emergence of the human sciences, and the formation of capitalist relations of production. Without the biopolitical project's necessary placement within a historical-social context, "bare life" becomes an abstraction whose complex conditions of emergence must remain as obscure as its political implications. Agamben tends to erase the historical difference between antiquity and the present, as well as the differences between the Middle Ages and modernity. Not only does he avoid the question of what biopolitics has to do with a political economy of life; he also suppresses the significance that gender has for his line of inquiry. He does not investigate to what extent the production of "bare life" is also a patriarchal project, one that codifies gender difference through a strict and dichotomous apportionment of nature and politics (cf. Deuber-Mankowsky 2002).

Agamben's books leave one with a surprising conclusion. Paradoxically, the author remains committed to precisely that juridical perspective and that binary code of law that he so vehemently criticizes and whose disastrous consequences he so convincingly illustrates. Agamben flattens the "ambiguous terrain" (1998, 143) of biopolitics by operating with a conception of the political that is as overloaded as it is reductionist. On the one hand, he conceives of the political as sovereign authority that recognizes nothing outside itself that would be more than an "exception." On the other hand, sovereignty utterly exhausts itself, in his interpretation, through the decisionist determination of the state of exception and the deadly exposure of "bare life."

Despite these criticisms, however, Agamben develops themes that often remain outside political theory. That is, he considers those themes that are "banned" from political reflection: life and death, health and sickness, the body and medicine. His theory serves to show that these problems are central to any consideration of the political and that the sphere of the political constitutes itself precisely through the exclusion of apparently apolitical "bare life." Likewise, *Homo Sacer* offers an analytical perspective that allows one to trace historical continuities and structural similarities between fascist or Stalinist regimes, on the one hand, and liberal democratic states, on the other. The political significance of Agamben's work lies in his making clear that it is not enough simply to expand the rights of those who hitherto have been without rights and therefore without protection. He insists that "the ways and the forms of a new politics" (1998, 187) are necessary; that is, a new political grammar is needed, one that annuls the difference between human and citizen altogether and transcends the legal conception that permanently presupposes and stabilizes the separation between political existence and natural being.[4]

5

Capitalism and the Living Multitude: Michael Hardt and Antonio Negri

IF FOR AGAMBEN biopolitics is marked by a catastrophic history that led to the Nazi extermination camps, it receives a very different treatment in yet another attempt at updating the concept. For the literary theorist Michael Hardt and the philosopher Antonio Negri, biopolitics does not stand for the overlapping of rule and exception but rather for a new stage of capitalism characterized by the disappearance of the borders between economics and politics, production and reproduction. In Hardt and Negri's cowritten works *Empire* (2000) and *Multitude: War and Democracy in the Age of Empire* (2004), they link their arguments to the Italian movement for workers' autonomy, ideas from classical political and legal theory, poststructuralist critiques centered on identity and the subject, and the Marxist tradition. The authors' goal is to combine these various theoretical sources and references in order to give a comprehensive account of contemporary processes of rule and, at the same time, the possibilities of political resistance.

The perspective of "biopolitical production" outlined in the two books resonated far beyond academic circles and university campuses and was the subject of passionate debate. This phenomenon was certainly helped by the fact that the antiglobalization movement received a boost at the beginning of the new millennium, as many activists searched for theoretical instruments with which to analyze international politics and the restructuring of contemporary capitalism. The writings of Hardt and Negri are also part of a larger network of research and dialogue. They had recourse to theses and positions that are, for example, developed in the contributions to the journal *Multitudes* and by the authors Judith Revel, Maurizio Lazzarato, and Paolo Virno.[1]

Imperial Rule and Immaterial Labor

In *Empire,* Hardt and Negri describe what they believe is an emerging new world order that is characterized by the tight interlocking of economic structures with juridico-political arrangements. "Empire" stands first of all for "a new form of sovereignty" (2000, xi) and a global system of domination. The authors argue that, faced with the development of trans- and supranational organizations, such as the United Nations or the European Union, and the growing importance of nongovernmental organizations, the regulatory power and authority of nation-states are losing their importance. Hardt and Negri also observe a shift away from traditional policies informed by constitutional guarantees to forms of intervention that follow the logic of a police state. These interventions function according to the definition of states of exception and operate in the name of higher ethical principles. In contrast to previous forms of sovereignty, the new imperial sovereignty knows no outside and has no center (ibid., 186–190). Rather, this new sovereignty is a network of self-referential and complementary political decision-making units that taken together establish a qualitatively different system of rule. Hardt and Negri see the economic dimension of Empire as a new stage of global capitalist production in which all states and regions of the world are integrated and connected. This foundational thesis of a boundless process of exploitation, however, does not refer solely to the configuration of the global market but also to a previously unreached depth of capitalist socialization. Today this includes not only the constitution of manpower but also the production of bodies, intellects, and affects.

Hardt and Negri argue that since the 1970s a decisive change in the modes of production has occurred. The paradigm of industrial capitalism, they write, has increasingly been replaced by "cognitive capitalism" (Negri 2008, 64). This form of capitalism is distinguished by an informatized, automated, networked, and globalized production process and leads to a decisive transformation in the working subject. Within this context, knowledge and creativity, language and emotion, are central to production and reproduction within society. According to Hardt and Negri, the informatization of production and its organization into networks make it increasingly difficult to maintain the division between individual and collective and intellectual and physical labor. The transformation of the production process leads to the dominance of a new form of socialized work, which the authors describe as "immaterial labor." The three most important aspects of immaterial labor are described by Hardt and Negri this way:

> The first is involved in an industrial production that has been informationalized and has incorporated communication technologies in a way that transforms the production process itself. . . . Second is the immaterial labor of analytical and symbolic tasks. Finally a third type of immaterial labor involves the production and manipulation of affects and requires (virtual or actual) human contact, labor in the bodily mode. (2000, 293)

The transformation of the mode of production includes a shift in the structures of exploitation. Capitalist exploitation operates today, the authors state, mainly through the absorption of the affective and intellectual capacity for work and the valorization of social forms of cooperation. Empire stands for the limitless mobilization of individual and collective powers in order to generate surplus value. All energies and spheres of life are subordinated to the law of accumulation: "There is nothing, no 'naked life,' no external standpoint, that can be posed outside this field permeated by money; nothing escapes money" (ibid., 32).

In this context, Hardt and Negri draw on Foucault's concept of biopolitics, but they submit it to an important revision. According to them, the creation of wealth in society "tends ever more toward what we will call biopolitical production, the production of social life itself, in which the economic, the political, and the cultural increasingly overlap and invest one another" (ibid., xiii). The authors describe biopower as "the *real subsumption* of society under capital" (ibid., 255; emphasis in original). They link the idea of an omnipresent and all-embracing biopower with ideas developed by the French philosopher Gilles Deleuze (1995). Deleuze argues in a brief essay that postwar Western societies have increasingly transformed themselves from "societies of discipline" into "societies of control." Control is exercised less through disciplinary institutions, such as schools, factories, and hospitals, than through the mobile and flexible networks of existence. Following Deleuze, Hardt and Negri conceive of biopolitics as a form of "control that extends throughout the depths of the consciousnesses and bodies of the population—and at the same time across the entirety of social relations" (2000, 24). It directs itself at social life as a whole, but it also includes the existence of individuals in the most intimate details of their everyday lives.

The authors critique Foucault as being too wedded to the paradigm of disciplinary power—an assessment that barely holds up in light of Foucault's analysis of liberal and neoliberal forms of government (cf. chapter 3 in this volume). Hardt and Negri impute to Foucault a "structural epistemology" (2000, 28) and a static notion of biopolitics. Whereas in their reading Foucault directs his attention excessively at top-down processes of power, they claim to look at the productive dynamic and creative potential of Empire. In order conceptually to mark these different foci, they distinguish in their subsequent book *Multitude* more strongly than before the terms "biopower" and "biopolitics": "Biopower stands above society, transcendent, as a sovereign authority and imposes its order. Biopolitical production, in contrast, is immanent to society and creates social relations and forms through collaborative forms of labor" (2004, 94–95; see also Negri 2008, 73–74).

The concept of "biopolitical production" stands here for a dual trend of capitalist socialization. It refers first of all to the dissolving of divisions between economics and politics that denotes a new stage of capitalist production. Here, in Hardt and Negri's view, the creation of "life" is no longer something both limited to the realm of reproduction and subordinated to the labor process; to the contrary, "life" now determines production itself. Consequently, the difference between reproduction and production increasingly loses significance. If biopower at one time stood for the reproduction of the relations of production and served to secure and preserve them, today it is an integral component of production. Empire is a "regime of biopower" (2000, 41) in which economic production and political constitution tend to overlap. The consequence of this is a wide-ranging convergence and parallelism between discourses and practices that have traditionally been separated from one another but that are now drawn into correlation:

> Production becomes indistinguishable from reproduction; productive forces merge with relations of production; constant capital tends to be constituted and represented within variable capital, in the brains, bodies, and cooperation of productive subjects. Social subjects are at the same time producers and products of this unitary machine. (2000, 385; see also 2004, 334–335)

Second, "biopolitical production" for Hardt and Negri also denotes a new relationship between nature and culture. It signifies a "civilization of nature" (2000, 187), "nature" here meaning everything previously external to the production process. Life itself becomes an object of technological intervention, and nature "has become capital, or at least has become subject to capital" (ibid., 32). Biological resources are the object of juridico-political regulation, while "natural" processes are opened up to commercial interests and potential industrial use. Nature thus becomes a part of economic discourse. Instead of being simply about exploiting nature, the discussion in the era of "sustainable" or "environmental capitalism" is about translating the biological and genetic diversity of nature into economic growth and opening it up to the development of profitable products and forms of life: "Previous stages of the industrial revolution introduced ma- chine-made consumer goods and then machine-made machines, but now we find ourselves confronted with machine-made raw materials and foodstuffs—in short, machine-made nature and machine-made culture" (ibid., 272).

Hardt and Negri see this double disappearance of demarcations as the transition from the modern to the postmodern. When econom- ics and politics and nature and culture converge, then there is no lon- ger an external standpoint of life or truth that might be opposed to Empire. This diagnosis grounds the perspective of immanence that underlies the authors' analysis. Empire creates the world into which it unfolds:

> Biopower is a form of power that regulates social life from its in- terior, following it, interpreting it, absorbing it, and rearticulating it. Power can achieve an effective command over the entire life of the population only when it becomes an integral, vital function that every individual embraces and reactivates of his or her own accord. (Ibid., 23–24)

To the degree that the imperial order not only rules over subjects but also generates them, exploits nature but also produces it, we are dealing with an "autopoietic machine" (ibid., 34) that reverts to im- manent justifications and rationales that it creates itself. Owing to this new biopolitical reality, it is no longer possible to uphold a dual perspective that operates on the basis of binary oppositions such as basis/superstructure, material reality/ideological veil, and being/ consciousness.

Multitude and the Paradoxes of Biopower

At this point, the description of an all-embracing and boundless system of rule reverts to a vision of resistance and liberation. While Hardt and Negri suggest that the whole of society will be subsumed under capital, they also couple this grim diagnosis with revolutionary hope. Biopolitics does not stand only for the constitution of social relationships that insert all individuals into a circulation of utility and value; it also prepares the ground for a new political subject. The biopolitical order that Hardt and Negri outline includes the material conditions for forms of associative cooperation that can abandon the structural constraints of capitalistic relations of production: "Empire creates a greater potential for revolution than did the modern regimes of power because it presents us, alongside the machine of command, with an alternative: the set of all the exploited and the subjugated, a multitude that is directly opposed to Empire, with no mediation between them" (2000, 393).

In opposition to imperial sovereignty, Hardt and Negri see the emergence of a "multitude." With this term the authors hark back to a concept derived from classical political theory, one that played a decisive role in the thinking of the early modern philosopher Baruch de Spinoza. "Multitude" describes the heterogeneous and creative whole of actors who move within power relations, without invoking a higher authority or an underlying identity. The multitude owes its formation to new conditions of production within a "globalized biopolitical machine" (ibid., 40). The "plural multitude of productive, creative subjectivities of globalization" (ibid., 60) is also the "living alternative that grows within Empire" (2004, xiii). The same competences, affects, and forms of interaction that are promoted by new structures of production and power also undermine them, in that they isolate themselves from monopolization and exploitation and arouse the desire for autonomous and egalitarian forms of life and relations of production. The authors outline the vision of a transformative force and a form of association that unites different kinds of social resistance and evades the political representation of peoples, nations, or class structures (2004, xiv–xv). The multitude represents a global countervailing force that signifies the possibility of liberation from domination and the prospect of new forms of life and work.

If biopower represents power over life, then it is precisely this life that constitutes the ground on which countervailing powers and forms of resistance are constituted. Biopolitics not only stands in opposition to biopower but also precedes it ontologically. Biopower is responsive to a lively and creative force that is exterior to it, which it seeks to regulate and shape, without being able to merge with it. Biopolitics refers here to the possibility of a new ontology that derives from the body and its forces. Such considerations gain support from Foucault's assessment of the conflictual field of biopolitics and the significance of resistance:

> If there was no resistance, there would be no power relations. Because it would simply be a matter of obedience. . . . So resistance comes first, and resistance remains superior to the forces of the process; power relations are obliged to change with the resistance. So I think that *resistance* is the main word, *the key word,* in this dynamic. (Foucault 1997a, 167; emphasis in original)

The militancy of the multitude remains committed to the insight that there is no standpoint outside of Empire. It "knows only an inside, a vital and ineluctable participation in the set of social structures, with no possibility of transcending them. This inside is the productive cooperation of mass intellectuality and affective networks, the productivity of postmodern biopolitics" (Hardt and Negri 2000, 413). The paradox of biopower, according to Hardt and Negri's reading, comes from the fact that the same tendencies and forces that maintain and preserve the system of rule are at the same time the ones that weaken and have the potential to overthrow it. It is precisely the universality and totality of this systematic nexus that makes it fragile and vulnerable: "Since in the imperial realm of bio-power production and life tend to coincide, class struggle has the potential to erupt across all the fields of life" (ibid., 403).

In this reading, Empire is a political picture puzzle. On the one hand, it represents a previously unknown control of life forces. It extends itself to all social relationships and penetrates the consciousness and the body of the individual. Since imperial rule is limitless and transgresses traditional demarcations between social fields and spheres of action, struggle and resistance are, on the other hand, always already economic, political, and cultural. Moreover, they have a productive and creative dimension. They not only set themselves against an established system of rule but also generate new forms of social life and political action: "they are biopolitical struggles, struggles over the form of life. They are constituent struggles, creating new public spaces and new forms of community" (ibid., 56).

Ontology and Immanence

The diagnosis of the present given in *Empire* and then expanded in *Multitude* has become the subject of a lively theoretical debate. If some critics see in Hardt and Negri's theses a "Communist Manifesto for the twenty-first century" that decisively enriches critiques of capitalism (Žižek 2001), others see in them a "sign of an apparently common intellectual flair for regression" (Lau 2002). A number of important objections have been formulated against the authors' thesis of an absolute structural rupture between modernism and postmodernism, imperialism and Empire. These objections expose the continuities and complementarities of various forms of exploitation and domination. It is by no means agreed that all "modern" differences and dualisms have a tendency to disappear or that they lose their social relevance, as Hardt and Negri predict. Binary codes, disciplinary techniques, and hierarchical structures continue to play a central role, as their substance and objects have proven themselves to be flexible and mobile. That Hardt and Negri do not consider the simultaneity and interconnectivity of heterogeneous technologies of power in their analysis but rather operate within a model of historical succession and systematic replacement shows that they are themselves wedded to a modern concept of the postmodern.

In the numerous reviews, commentaries, and critiques of Hardt and Negri's two works, comparatively little was written about the concept of biopolitics. But it is especially in this area that the problems of Hardt and Negri's argumentation are clearest. As important as it is for the authors to discuss the "revolutionary discovery of the plane of immanence" (2000, 70), they also neglect to sustain and then implement this theoretical perspective. While the authors demonstrate the impossibility of an "external position" within Empire, their reference to "life" breaks with the principle of immanence. "Life" in this instance is not, as it is with Foucault (1970), configured as a social construct or as an element of historical knowledge; rather, it functions as an original and transhistorical entity. The ontological conception of biopolitics proposed by Hardt and Negri is on the one hand so comprehensive that it remains unclear in what way it might be circumscribed and how it relates to other forms of political and social action. On the other hand, it allows for the implementation of a well-considered choreography that consistently counterposes two principles, instead of analyzing them on the "plane of immanence," which is what the authors demand. The vital and autonomous multitude struggles against the unproductive, parasitical, and destructive Empire.

Hardt and Negri's diagnosis of the rule of Empire corresponds with a glorification of the multitude. The multitude alone is productive and positive, whereas Empire is controlling and restrictive. To Hardt and Negri, the "specificity of corruption today is instead the rupture of the community of singular bodies and the impediment to its action—a rupture of the productive biopolitical community and an impediment to its life" (ibid., 392). It is questionable, however, whether production and regulation can be so cleanly separated: Is not every instance of production always already a kind of regulated production? Why does Empire produce only that which is negative, and the multitude something positive? Are emotions or desire not always already a part of Empire, reproducing and stabilizing it? Instead of conceiving of the relation between Empire and multitude as one between two ontological entities, it would be more appropriate to analyze a (biopolitical) relation of production that contains both poles within it.

Hardt and Negri do not limit themselves to tracing the historical emergence of a new political figure. They tend rather to anchor the multitude ontologically. Negri discusses, for example, "biodesire," which is contrasted with biopower: "The desire for life, the strength and wealth of desire, are the only things that we can oppose to power, which needs to place limitations upon biodesire" (Negri 2004, 65). There is a danger that the ontological rendering of biopolitics, quite contrary to the intentions of the authors, has the effect of depoliticizing their work, when they conceive of the multitude per se as an egalitarian and progressive force that is invested with a radical-democratic goal. Instead of contributing to social mobilization, this way of thinking could, on the contrary, leave the impression that political struggles are nothing other than incarnations of abstract ontological principles that almost automatically proceed without the engagement, intention, or affect of concrete actors (Saar 2007, 818).

The contrast of Empire and multitude and the antagonism between a productive and creative biopolitics from below and a parasitical and vampiric biopolitics from above lead to a theoretical dead end. The authors do not do justice to the complexity of the problem of Empire's political constitution. This has less to do with the hindering of activity, its limitation or canalization, than it does with the incitement to specific (and in this respect selective) activities. It has less to do with the contrast between production and destruction than with the promotion of a destructive production. Seen this way, it is not about establishing the difference between production and nonproduction or imputing the driving forces of "biodesire," as Hardt and Negri suggest. It is rather about the invention of a production that has other goals and about the fostering of a desire for alternative forms of life that are autonomous and egalitarian.

6

The Disappearance and
Transformation of Politics

THERE CAN BE no doubt that the writings of Giorgio Agamben and the works of Michael Hardt and Antonio Negri are the most prominent contributions to the debates concerning the further development and actualization of Foucauldian biopolitics. However, numerous other attempts to grapple with the concept have been made. One can generalize by saying that there are two primary threads by and through which the term has been adopted. The first, which is introduced in this chapter, is to be found above all in philosophy and social and political theory. This area of inquiry concentrates on the mode of the political: How does biopolitics function, and what counterforces does it mobilize? How does it differentiate itself analytically and historically from other eras and from other political formations? The second domain in which biopolitics plays an important role is the subject of the next chapter. It originates in science and technology studies, medical sociology, and anthropology, as well as in feminist theory and gender studies. The main focus here lies in the substance of life. If as a consequence of bioscientific innovations the living body is regarded today less as an organic substratum than as molecular software that can be read and rewritten, then the question as to the foundations, means, and ends of biopolitics needs to be posed in a different manner.

Central to this chapter are three important theoretical approaches that have greatly influenced these debates and that deal with the question of the relationship between biopolitics and "classical" politics in different ways. The philosophers Agnes Heller and Ferenc Fehér understand biopolitics as a regression of the political, since, according to them, its immediate relationship to the body has a totalitarian tendency that threatens freedom. By contrast, the sociologist Anthony Giddens presents his concept of life politics as advancing and adding to traditional forms of political articulation and representation. A third position is taken by the medical anthropologist Didier Fassin. His term "biolegitimacy" stands neither for the negation of established political forms nor for their perpetuation; rather, it traces a fundamental political realignment whereby the sick or injured body is assigned a central political meaning.

Body Politics

The political philosophers Ferenc Fehér and Agnes Heller published their book *Biopolitics* in 1994. It provides a view of this field of inquiry that can be clearly distinguished from the naturalist and politicist theoretical tradition, as well as from the interpretation of biopolitics that Foucault shaped. Life appears here neither as foundation nor as object but as a counterprogram to politics. Fehér and Heller view the increased social meaning of the body as political regression and sharply delimit biopolitics from "traditional modern politics" (Fehér and Heller 1994, 38; see also Heller 1996). This caesura nevertheless appears to be different from the one that Foucault allows for between sovereign power and biopower. To be sure, these authors refer to Foucault's definition of biopolitics by distinguishing between individual discipline and regulation of the collective body (Fehér and Heller 1994, 10). However, they view biopolitics not as a product of modernity but rather as its antithesis.

The crux of their analysis relies on the academic debates and the media discussions about health, the environment, gender, and race that occurred in the United States in the 1990s. The authors situate these "biopolitical" themes within a political theory of modernity.

Fehér and Heller view biopolitics as a "politics of the Body" that emerged with the modern era and whose significance is still growing (1994, 17; Heller 1996, 3). The book follows the historical metamorphosis of this form of politics and investigates its consequences, from health to the "race question." The authors are concerned with the "totalitarian venom" that threatens discussions related to biopolitical problems (Fehér and Heller 1994, 27). Their critique is aimed at new social movements, above all at feminism and the peace movement, but also directs itself against the "postmodern" academic-cultural Left.

Fehér and Heller refer to Nazism as "an early experiment with biopolitics" (ibid., 21); this is distinguished from contemporary biopolitics, which they see as having integrated itself into democratic processes. Fehér and Heller consider its "intellectual mentor" to be postwar French philosophy (ibid., 51), which they argue is marked by skepticism toward universal principles, an emphatic demand for "difference," and the privileging of the aesthetic with respect to ethical questions (ibid., 51–57). The authors view the conflict between freedom and life as central to the understanding of contemporary biopolitics. The "début" of the new biopolitics, they argue, took place during the peace movements of the 1980s, which inaccurately assessed the aggressiveness of Soviet politics and consequently valued life and survival higher than freedom from tyranny and oppression (ibid., 22). Biopolitical movements and positions as characterized by these authors are marked by a tendency to value life over freedom. Their warning, therefore, applies to that "point at which . . .—in the name of the integrity of The Body—freedom is sacrificed" (ibid., 104).

In the arguments of Fehér and Heller, the verdict on biopolitics addresses a heterogeneous field of social actors and political interests. These comprise not only the peace movement but also feminist positions, health and environmental groups, and antichoice as well as prochoice advocates in the abortion rights debate. Common to all these cases is the fact that the "infatuation with the task of releasing the Body from its bonds was so feverish" that the resultant problems were either not seen or not seen well enough (ibid., 9). The authors' analysis contains a range of trenchant observations and critical arguments but remains on the whole astonishingly one-sided and vague. The main reason for this is that Fehér and Heller substitute a polemical strategy for categorical clarity and analytical precision. Nonetheless, they point out important problematic "biologizations" of social conditions and rightly criticize the morally charged nature of the health discourse (cf. ibid., 67–68).

Their analysis of biopolitics, however, is reductionist in two ways (cf. Saretzki 1996). First, Fehér and Heller treat biopolitical themes as an antithesis between freedom and totalitarianism. This viewpoint may result from the authors' life experiences. They are both former dissidents who lived in Hungary under a socialist dictatorship and emigrated to the United States in the 1970s. However, the basic choice between life and freedom does not do justice to the complexity of biopolitical questions in the political and historical context that has changed since then. Fehér and Heller systematically underestimate the seriousness of the problems that they examine and that are found in liberal democratic societies—such problems as health, gender, the environment, and ethnicity. Within their theoretical system, issues such as distributive justice, participation, and solidarity are not even raised, or they are immediately interpreted as manifestations of totalitarian domination. Fehér and Heller also fail to appreciate that many of the questions they treat elude clear analytical ordering or normative valuation.

Second, Fehér and Heller conceive the scope and dimensions of biopolitics too narrowly. They see biopolitics as, in principle, the alternative model to classical politics—a kind of antipolitics. Upon closer inspection, however, it becomes apparent that Fehér and Heller use the term "biopolitics" very restrictively. It designates a specific form of political action that is calibrated solely to corporeal themes. Central to their analysis, then, are not so much life processes in a comprehensive sense, which would also include environmental or nonhuman themes, but forms of political action that focus exclusively on the human body. Many biopolitical conflicts that are initiated by questions related to animal rights or the patenting of life can hardly be conceived of as being part of the "liberation" of the human body.

Yet Fehér and Heller restrict not only the empirical scope of biopolitics; the way they analyze biopolitical phenomena is also extremely selective and one-sided. They primarily discuss biopolitical problems and themes from the perspective of ideology critique, generally denying them any material content. Whether it is a reference to the hazards of passive smoking or an analysis of the depletion of natural resources or of species extinction, the authors approach these debates solely with regard to possible "instrumentalization" and "functionalization" while neglecting the materiality of these problems.

Fehér and Heller ask questions that are certainly worth pursuing. One example is their intuition that references to life and its improvement, which are voiced in the demands of many social movements, restrict freedoms and lead to new forms of exclusion and oppression. On the whole, however, their analysis is too schematic. Even if one accepts the perspective they adopt (the dualism between freedom and life) as appropriate for an analysis of contemporary biopolitics, it still remains unclear how Fehér and Heller themselves would resolve the contradiction they perceive—that is, how precisely they want to mediate between these two values. A reading of their work suggests, rather, that they regard this as a pseudoproblem, since in each case freedom has priority over life. As a result, their analysis remains situated in a relatively simple theoretical framework that can ultimately be reduced to a series of binaries: modernity is opposed to postmodernity, traditional politics to biopolitics, and freedom to life.

Life Politics

In the 1990s, when Heller and Fehér were formulating their critique of biopolitics, the influential sociologist Anthony Giddens developed his concept of "life politics" (1991, 209–231; cf. also 1990). Giddens does not make explicit reference to Foucault and his understanding of "biopolitics." As a frame of reference, he prefers to employ his own theory of reflexive modernization, which exhibits a range of similarities, but also important differences, to Ulrich Beck's concept of "second modernity."

Giddens begins his analysis with the observation that in the closing decades of the 20th century modernity entered a new stage, namely, the late modern. This new phase does not represent the end of the modern, as postmodern diagnoses might suggest, but rather its advancement and radicalization. Giddens's starting point is the problem of "ontological security" under the conditions of modernity. He conceives of insecurity and uncertainty in relation to social realities not as a premodern residue but, on the contrary, as an achievement of modernity. Modernity, he argues, cultivated and institutionalized both doubts about received tradition and skepticism with regard to fundamental truths, by opening them up for rational argumentation and democratic negotiation, thereby providing the ground for establishing new traditions and certitudes.

Central to Giddens's argument is the concept of reflexivity. Modernity, he says, is characterized by the perpetual revision of convention, which in principle encompasses all areas of life and fields of action. Giddens argues that social practices are constantly monitored and changed in the light of new knowledge about these practices. In this respect, the knowledge of the participants is itself an element of social practice. However, the reflexivity of social life in modern society has its price: it erodes the notion of a stable and ultimate knowledge, since the principle of reflexivity must be applied to itself. The result is that both the content and the production of knowledge are treated as provisional and revisable.

Late modernity accentuates this problem. Giddens sees the increase in forms of knowledge and the possibilities of intervention as defining late modernity; they are expressed in institutional reflexivity and in a reflexive conception of the body and the self. In place of predefined concepts of life and rigid social roles, a culture of negotiation, choice, and decision-making increasingly comes to the fore. Ways of living also become freely configurable to a degree unknown before.

With the transition from modernity to late modernity, Giddens also sees a fundamental change in the political. He argues that modernity is for the most part marked by a political form he calls "emancipatory politics," a term he uses to refer to practices that have as their goals liberation from social and political coercion and the overcoming of illegitimate rule. Emancipatory politics works against three symptoms of power—exploitation, inequality, and suppression—and attempts in turn to anchor ideas of justice, equality, and participation in social institutions. One of its purposes is to free underprivileged groups from their condition or at least to reduce the imbalance of power between collectives. Although the concerns of emancipatory politics have decisively advanced the project of modernity, today one can observe a new type of politics that represents a fundamentally different way of understanding. Giddens describes this new form of politics as "life politics" and understands this to be "radical engagements which seek to further the possibilities of a fulfilling and satisfying life for all" (1990, 156). Whereas "emancipatory politics" is a politics of life chances, "life politics" pursues a politics of lifestyle. If the former harks back to notions of justice and equality, the latter is driven by the quest for self-actualization and self-identity. It is founded less on political programs than on personal ethics. For Giddens, the protagonists of this novel form of politics are members of new social movements, particularly the feminist movement, which unite the personal with the political.

According to Giddens, two complementary processes have made this form of politics possible. First, body and self are increasingly viewed as flexible and alterable as well as subject to processes of knowledge formation. The body is no longer conceived of as a fixed physiological-biological entity but is seen as involved in the project of reflexive modernity. In a posttraditional society, the individual body is a central point of reference for the formation of social identity. Increasingly available biotechnological and medical interventions also play a decisive role and lead, according to Giddens, to the "end of nature" (1991, 224). Nature has ceased to be our destiny. Where once there was fate, Giddens now sees scope for transformation and intervention and the necessity to decide on the different options available.

As an example, Giddens invokes reproductive technologies, which make available a multiplicity of choices, divide sexuality from reproduction, and render obsolete traditional notions of fertility and parenthood. For Giddens, reproduction is no longer a matter of accident or destiny but rather is an expression of personal preference and choice. The "disappearance of nature" and the emergence of new options impinges, however, not only on the question of reproduction but also on bodily appearance and sexual orientation, both of which are viewed as increasingly changeable, correctable, and open to intervention.

The consequences of biotechnological and medical innovations for the formation of individual identity constitute only one aspect of "life politics." Giddens also emphasizes personal decisions and everyday practices that strongly influence macroprocesses and global phenomena. For example, reproductive decisions connect individual choices and options to the survival of the human species. Similarly, there are connections between lifestyle, consumer options, and ecological questions.

At its core, life politics touches on the question "How should we live?" This is a question that Giddens feels must be answered in a posttraditional context. On the level of everyday behavior and private life, but also in the domain of collective practices, ethical questions need to be raised and openly debated. Abstract statistics, probability calculations, and risk assessments must be translated into existential judgments. Moral dilemmas replace clear certitudes and scientific explanations. In this manner, life politics contributes to a remoralization of social life and brings about a new sensitivity to questions that in modern institutions have until now been marginalized or suppressed (1991, 223)

Worthy of consideration though Giddens's arguments are, his conception of life politics is not, on the whole, convincing. The reasons for this lie above all in the fact that his differentiation between emancipatory politics and life politics is not sufficiently elaborated and remains to a large extent diffuse, just as his separation of the late modern from the classic modern does. Giddens tries on the one hand to reinforce the continuity of modernity, but he simultaneously recognizes a decisive caesura within the project of the modern. His argument constantly shifts backward and forward between these two positions, which results in two problems.

First, many of the phenomena that Giddens takes to be typical of late modernity are also apparent in earlier eras. The rupture he identifies between modernity and late modernity is itself characteristic of modernity. Giddens's concept of the modern is one-dimensional and reductive. It is to a large extent limited to principles of individual autonomy, self-determination, and freedom of action, while it neglects dimensions of aesthetic and cultural modernity. Thus, Giddens disregards those voices that, from within the project of modernity, have drawn attention to its limits and contradictions in order to criticize reification, alienation, and repression in the name of modernity.

Second, Giddens's concept of life politics is curiously apolitical because it completely lacks the subversive and resistant moments that transcend the modern order. Unlike postmodern "identity politics," from which Giddens seeks to distance himself and through which sexual, religious, and ethnic minorities claim difference and challenge the universalism of modernity, late-modern life politics begins to resemble a "politics of lifestyle" (1991, 214) or a *politics of self-actualisation*" (1990, 156; emphasis in original). Giddens says nothing about how precisely individual self-actualization is linked to collective decision-making processes and what kinds of representation and modes of articulation would be necessary for that to happen. Given this apparently very comprehensive, but in fact meaningless, conception of politics, it is hardly surprising that Giddens deals only marginally with central dimensions of the politicization of personal existence. His discussion of life politics concentrates almost exclusively on forms of knowledge and the possibilities of intervention with respect to human nature. In contrast to these concerns, he is less interested in the interplay of social relationships and environmental problems.

Although Giddens explains that life-political interests do not remove or suppress emancipatory political concerns, he does implicitly work with a phase model. He tends to connect life politics with a specific societal stage of development and to limit its meaning to industrialized and "modernized" societies. It is questionable, however, whether his distinction between questions of redistribution and inequality, on the one hand, and those of identity and recognition, on the other, is tenable. It is not possible historically or systematically to isolate emancipatory politics from life politics or to define them against each other (Flitner and Heins 2002, 334–337).[1]

Biolegitimacy

Less well known than the works of Fehér, Heller, and Giddens is medical anthropologist Didier Fassin's concept of "biolegitimacy." In his books and articles of recent years he has shown that biopolitical phenomena always have a moral dimension, which means that every analysis of the politics of life must also take account of the underlying moral economy. Fassin understands morality to be not the establishment of values or the distinction of right from wrong but rather the development of norms in a given historical and geographical context, which is accessible to ethnological investigation. Fassin stresses that the inclusion of the moral dimension does not replace political analysis; rather, it expands and deepens it. His central question is, what are the systems of values and normative choices that guide the politics of life? (Fassin 2006).

Fassin distinguishes between two aspects of this moral dimension. First, matters of life and longevity, health and illness, are not to be separated from those of social inequality. That a thirty-five-year-old, unskilled worker in France can expect on average to live a life nine years shorter than that of an engineer or teacher of the same age, that the statistical life expectancy in Uganda is half as high as it is in Japan—these data reflect collective choices and normative preferences in a given society or on a global level. These decision-making processes remain, according to Fassin, mostly implicit, since governments are only rarely prepared to declare publicly that they allow some people to live shorter lives than others or even that some people are sacrificed for others.

A second moral dimension of biopolitics for Fassin goes beyond this distinction between the differing life expectancies and qualities of life for the rich and the poor and for the rulers and the ruled. Instead of measuring life comparatively in quantitative or qualitative terms, this second form of moral reflection embraces the concept of life itself. In this case, Fassin turns to Agamben's distinction between bare life (zoé) and political existence (bíos), though he gives these terms a definition that significantly distinguishes them from Agamben's usage.

In contrast to Agamben, Fassin states that the biopolitical relationship between body and state does not take the form of a violent prohibition or ban. He sees rather a subtle government of bodies at work, one that organizes itself around health and corporeal integrity as central values. "Bare life" appears from this perspective as the vector of a "biolegitimacy" that precludes recourse to violence. Whereas Agamben diagnoses a "separation between humanitarianism and politics" (1998, 133), humanitarianism is the quintessential form of biopolitics for Fassin. Humanitarianism is not a closed social field of action that is defined and administered by large NGOs but rather a moral principle that grants human life absolute priority. According to Fassin, it is more and more apparent in the social domain that the body functions as the final authority on political legitimacy (2006).

Fassin illustrates his thesis by looking at French refugee policies of the past twenty years. Throughout the 1990s, one can observe two opposed but, in Fassin's view, complementary trends. On the one hand, the number of recognized asylum seekers sank to a sixth of the 1990 figure, above all because of an increasingly restrictive interpretation of the right to asylum. On the other hand, the number of refugees who received a temporary right of residence because they suffered from a disease untreatable in their home countries rose sevenfold during the same period. Fassin argues that these contrary developments show a systematic shift in social definitions of legitimacy.

The growing recognition and acknowledgment of the life of a human being who suffers from an illness displaces the recognition of the life of a citizen who has experienced violence, often resulting from political agitation. In place of political life that confronts a legal-administrative order to reconstruct the history of a persecution, we find biological life that documents a history of illness against the background of medical knowledge. The right to life has increasingly moved from the political arena to the humanitarian one. According to Fassin, it is now apparently more acceptable to reject an application for asylum as unfounded than to reject a medical report that recommends temporary residency for medical reasons (2006, 2001).

The emergence of biolegitimacy—the recognition of biological life as the highest value—in no way limits itself to refugee policies. Fassin exposes the logic of humanitarianism in many social fields. The connection to health and bodily integrity has also led to a realignment of social-political programs and measures. People once considered underprivileged or deviant would today increasingly be considered suffering bodies in need of medical care. Thus, heroin addicts have come to be viewed as possible victims of deadly infections rather than dangerous delinquents or threats to society. Similarly, there is a growing recognition of the psychic and physical suffering caused by material poverty and social exclusion.

Fassin's assessment of the social development resulting in "biolegitimacy," in which humanitarian logic is the highest ethical ideal, is ambivalent. To be sure, he welcomes what he sees as the tendency for punishment to be replaced by care, and surveillance by compassion. However, he also believes this has led to a mitigation of and a reformulation of political problems as moral and medical ones. Social suffering thus mixes with bodily suffering, and the boundary between the social and the medical dissolves. Fassin states that in order to analyze contemporary societies it is necessary not only to consider biopower as power over life but also to view biolegitimacy as the legitimacy of life, since government operates not so much on the body as through it (2005).

Body politics, life politics, and biolegitimacy—these interpretive lines admittedly represent only a small portion of the philosophy and social theory devoted to biopolitics. Two further attempts at updating the concept should at least be mentioned here. *Bíos: Biopolitics and Philosophy* (2008) is the only book by the Italian philosopher Roberto Esposito thus far translated into English. It is the last volume of a trilogy and marks a highpoint of philosophical reflection on the "enigma of biopolitics" (ibid., 13).[2] Esposito's central thesis is that modern occidental political thinking is dominated by the "paradigm of immunization" (ibid., 45). He shows, via a reconstruction of political theory since Thomas Hobbes, that the modern concepts of security, property, and freedom can be understood only within a logic of immunity. Characteristic of this logic is an inner connection between life and politics, in which immunity protects and promotes life while also limiting life's expansive and productive power. Central to political action and thinking is the safeguarding and preservation of life. This goal ultimately leads to (self-)destructive results. In other words, to the extent that the logic of immunity secures and preserves life, it also negates the singularity of life processes and reduces them to a biological existence. This "immunitary logic" leads from the maintenance of life to a negative form of protecting it and finally to the negation of life (ibid., 56).

The paradigm of immunity allows the two opposing dimensions of biopolitics (advancement and development of life, on the one hand, and its destruction and elimination, on the other) to be conceived of as two constitutive aspects of a common problematic. Esposito sees Nazi racial and extermination policies as the most extreme form of immunitarian rationality, in which life politics is entirely enveloped in a negative politics of death (thanatopolitics). Esposito, along with Agamben and Foucault, stresses that Nazism stands in a continuum with modern political thinking and action. However, in contrast to the other two philosophers, he sees the specifics of Nazism neither in the rearticulation of sovereignty nor in the supremacy of the state of exception. Esposito highlights instead the medical-eugenic goals of Nazism and the programmatic importance of the fight against illness, degeneration, and death. The immunitarian project of promoting life leads ultimately to the death camps:

> The disease against which the Nazis fight to the death is none other than death itself. What they want to kill in the Jew and in all human types like them isn't life, but the presence in life of death: a life that is already dead because it is marked hereditarily by an original and irremediable deformation; the contagion of the German people by a part of life inhabited and oppressed by death. . . . In this case, death became both the object and the instrument of the cure, the sickness and its remedy. (Ibid., 137–138)

As a countermodel to this "thanatopolitics"—which did not disappear with the end of Nazism but continues to characterize the current era (cf. ibid., 3–7; Campbell 2008)—Esposito presents an "affirmative biopolitics," whose main point of reference is to incomplete and open individual and collective bodies. These bodies defend themselves against attempts at identification, unification, and closure and articulate an immanent normativity of life that opposes the external domination of life processes. This vision of an affirmative biopolitics should be "capable of overturning the Nazi politics of death in a politics that is no longer over life but *of* life" (Esposito 2008, 11, emphasis in original). It would substitute a new concept of community for the self-destructive logic of immunity. This new concept recognizes the constitutive vulnerability, openness, and finitude of individual bodies and the collective body as the essential foundation of community—rather than consistently viewing them as a danger to be repelled.

The French medical anthropologist Dominique Memmi adopts a very different starting point. In *Faire vivre et laisser mourir: Le gouvernement contemporain de la naissance et de la mort* (2003a), she observes a shift in biopolitical mechanisms during the past thirty years. Memmi stresses that biopolitical processes limit themselves less and less to the forms of discipline and population regulation which Foucault investigates in his work. On the contrary, citizens themselves are granted the right to make life and let die. This applies above all to questions pertaining to the beginning and end of life. From the deployment of reproductive technologies—for example, in-vitro fertilization—to the decriminalization of abortion ("let live or prevent from living") to assisted death in palliative care ("letting oneself die") or consciously induced death through assisted suicide ("making oneself die")—all these cases, argues Memmi, have to do with decisions that are increasingly the responsibility of individuals. She points out that "self-determination" is a central feature of contemporary biopolitics. The traditional care of the state over individual bodies and the population's health as a whole is today absorbed into forms of self-care. This does not, however, signify a simple growth in individual autonomy. Rather, a new type of social control is established whereby only those decisions about the body that conform to social expectations and norms are considered rational, prudent, or responsible (Memmi 2003b).

7

The End and Reinvention of Nature

A SECOND SIGNIFICANT line of reception linked to Foucault's concept of biopolitics focuses on the manner in which new scientific knowledge and the development of biotechnologies increase the control of life processes and decisively alter the concept of life itself. The common starting point for work in this field is the observation that the image of a natural origin of all living organisms is gradually being replaced by the idea of an artificial plurality of life forms, which resemble technical artifacts more than they do natural entities. The redefinition of life as text by geneticists, advancements in biomedicine that range from brain scans to DNA analysis, transplant medicine, and reproductive technologies—to name but a few innovations—represent a rupture with the perception of an integral body. The body is increasingly seen not as an organic substratum but as molecular software that can be read and rewritten.

In light of these developments, a series of works has proposed a critical review of, and amendment to, the Foucauldian concept of biopolitics. These works concentrate less on the transformation of politics and more on the "reinvention of nature" (Haraway 1991). Since it is not possible to present this vast literature in all its diversity here, this chapter considers three core themes of the discussion.

The first group of studies accentuates the extension and relocation of biopolitical interventions. From this perspective, biotechnological practices increasingly include the body's interior as a new space for intervention. In addition, they create a new relationship between life and death and dissolve the epistemic and normative boundaries between the human and the nonhuman. Next, I briefly introduce the thesis of anthropologist Paul Rabinow, according to which new socialities and forms of political activism emerge on the basis of biological knowledge. There then follows a discussion of sociologist Nikolas Rose's concept of ethopolitics.

Molecular Politics, Thanatopolitics, Anthropopolitics

Foucault's concept of biopolitics remains bound to the notion of an integral body. His analyses of disciplinary technologies which are directed at the body, in order to form and fragment it, are based on the idea of a closed and delimited body. By contrast, biotechnology and biomedicine allow for the body's dismantling and recombination to an extent that Foucault did not anticipate. A range of authors have therefore pointed out the limits of Foucault's concept of biopolitics. Michael Dillon and Julian Reid (2001) suggest that molecularization and digitalization characterize a "recombinant biopolitics," which operates within and beyond the body's boundaries. According to another thesis, advancements in the biosciences have established a new level of intervention below the classic biopolitical poles of "individual" and "population." Michael J. Flower and Deborah Heath (1993) argue that a "molecular politics" has emerged that no longer proffers an anatomical view of individuals but rather presents a genetic one which situates the individual in the "gene pool."

Donna Haraway, Hans-Jörg Rheinberger, and many other theorists of science have drawn attention to the fact that with regard to these processes it is not just a matter of enhancing preexisting technologies and instruments. On the contrary, genetic engineering clearly distinguishes itself from traditional forms of bioscientific and medical intervention since it aims to "reprogram" metabolic processes, not merely to modify them. Central to this political epistemology of life is no longer control of external nature but rather the transformation of inner nature. As a consequence, biology is conceived of no longer as a science of discovery that registers and documents life processes but rather as a science of transformation that creates life and actively changes living organisms (Haraway 1991; Rheinberger 2000; Clarke et al. 2003).

Marcela Iacub, Sarah Franklin, Margaret Lock, Lori B. Andrews, Dorothy Nelkin, and others point to a further aspect of the biopolitical problematic, showing that enhanced access to the body also creates a new relationship between life and death. In two respects, life and death today are more closely linked than Foucault assumed. To start with, "human material" transcends the living person. The person who dies today is not really dead. He or she lives on, at least potentially. Or more precisely, parts of a human being—his or her cells or organs, blood, bone marrow, and so on—can continue to exist in the bodies of other people, whose quality of life they improve or who are spared death through their incorporation. The organic materials of life are not subordinate to the same biological rhythms as the body is. These materials can be stored as information in biobanks or cultivated in stem cell lines. Death can be part of a productive circuit and used to improve and extend life. The death of one person may guarantee the life and survival of another.

Death has also become flexible and compartmentalized. The concept of "brain death" and the development of reanimation technologies, as well as the splitting of death into different regions of the body and moments in time, has allowed for the growth and spread of transplantation medicine. Today, it is not so much state sovereignty as medical-administrative authorities who decide on matters of life and death. They define what human life is and when it begins and ends. In an entirely new sense, "thanatopolitics" is an integral part of biopolitics (Andrews and Nelkin 2001; Iacub 2001; Franklin and Lock 2003).

A third critique of Foucauldian biopolitics is that it is exclusively oriented to human individuals and populations. As Paul Rutherford (1999) correctly argues, one cannot determine within such a narrow conceptual framework how ecological problems and the environmental discourse mesh with the (re)production of the human species. He suggests an expansion of the semantic field that would allow the concept of biopolitics to stand for the administration and control of the conditions of life. Yet a further problem reveals itself, namely, that Foucault conceived of agency as a human quality, so that only humans as social actors are taken into account. Gesa Lindemann and Bruno Latour have convincingly and from different perspectives criticized this anthropocentric curtailment of the biopolitical problematic. Lindemann (2002) suggests, with reference to Helmuth Plessner's work, a "reflexive anthropology" which asks who is empirically included within the circle of social persons. Bruno Latour (1993) moves in a similar direction with his demand for a "symmetrical anthropology" that conceives of both human and nonhuman entities as capable of action.

These and other theoretical contributions are opening up a new field of research that makes it possible to investigate which entities, under what conditions, can become members of society and which cannot: biopolitics as anthropolitics.

Biosociality

In a widely read essay, Paul Rabinow (1992) introduces the concept of biosociality as an extension of Foucault's biopolitical problematic. Rabinow sees a new articulation of the two poles Foucault identified (body and population) emerging from the Human Genome Project and the biotechnological innovations linked to it. Rabinow believes a postdisciplinary order has emerged, one in which the strict division between nature and culture has been overcome and in which a different relationship to life processes is developing (ibid., 234). In this context, it is not enough to describe the "new genetics" in terms derived from previous eras. Rabinow holds that it is no longer accurate to speak of the biologization of the social or the translation of social projects into biological terminology (in the light of the well-known models of sociobiology or social Darwinism); he argues that we are instead confronted with a new understanding of social relationships through biological categories:

> In the future this new genetics will cease to be a metaphor for modern society and will become instead a circulation network of identity terms and restriction loci around which and through which a truly new type of autoproduction will emerge, which I call "biosociality." If sociobiology is culture constructed on the basis of a metaphor of nature, then in biosociality, nature will be modeled on culture understood as practice. (Ibid., 241)

Rabinow is especially interested in how, within the context of growing knowledge about genetic diseases and genetic risks, new individual and collective identities emerge. It is to be expected, he argues, that to the extent that genetic information spreads and is popularized, people will describe themselves and others in bioscientific and genetic terminology, as biomedical vocabulary seeps into everyday language. Just as people today describe themselves in terms of low blood pressure or high cholesterol, people in the future may define themselves in terms of their elevated genetic risk for this or that illness, their genetically conditioned low tolerance for alcohol, or their inherited predisposition to breast cancer or depression.

Yet Rabinow's thesis goes even further. Technical innovations and scientific classification systems create, he postulates, the material conditions for new forms of socialization, representational models, and identity politics, whereby knowledge about specific bodily properties and genetic characteristics decisively determine the relationship of the individual to her- or himself and to others:

> [T]here will be groups formed around the chromosome 17, locus 16,256, site 654,376 allele variant with a guanine substitution. These groups will have medical specialists, laboratories, narratives,

traditions, and a heavy panoply of pastoral keepers to help them
experience, share, intervene in, and "understand" their fate. (Ibid.,
244)

According to Rabinow, self-help groups and patient organizations
are not passive recipients of medical care or the objects of scientific
research interests. On the contrary, the experience of illness forms
the basis of a field of diverse social activities. Groups of people with
a given illness and their families work *with* medical experts. They col-
lect donations in order to promote research targeted at their needs,
and they build networks of communication that range from regular
group meetings to exchanging stories of their own experience of illness, from running their own publications to creating sources of information on the Internet (cf. Rabinow 1999).[1]

The spread of bioscientific and medical knowledge, however, not
only leads to new forms of community and collective identity. It also
results in a demand for rights based on biological anomalies and in
hitherto unknown forms of political activism. In the Anglo-Ameri-
can world, these new modes of articulation and representation are
defined and discussed in such terms as "biological" or "genetic citizenship" (Petryna 2002; Heath, Rapp, and Taussig 2004; Rose and
Novas 2005). What these concepts have in common is the idea of a
systematic connection between biomedical knowledge, concepts of
identity and selfhood, and modes of political articulation. From this
perspective, patient organizations, self-help groups, and family associations represent new collective subjects that remove the borders
between laypeople and experts, between active researchers and the
passive beneficiaries of technological progress.

At least three arenas of political activism organized around shared
biological attributes can be distinguished. First, self-help groups, patient organizations, and family associations work as lobbyists in order to increase public interest in their concerns and to attract state

funding for research projects related to their respective causes. Their goal is to sensitize the public to the concerns of the sick and their suffering and to influence policy-makers.

A second arena of political activism is the struggle against material or ideological restrictions to gain access to medical technologies and to bioscientific knowledge. Self-help groups and patient organizations fight restrictive or exclusive concepts of intellectual property in the domain of biomedical and genetic research. They also direct their resources against the use of genetic knowledge solely for commercial uses, which can lead to limitations on further research and to increases in the cost of the development and dissemination of diagnostic and therapeutic devices. A third field of engagement on the part of self-help groups and patient organizations is their participation in ethics committees and parliamentary deliberations, as well as the drafting of guidelines for the regulation of technological procedures (Rabeharisoa and Callon 1999; Rabinow 1999; Heath, Rapp, and Taussig 2004; Rose and Novas 2005).

Until now, only a few studies have tracked this "biopolitics from below" and empirically examined the relationship between collective forms of action and group identities of patient and family organizations. For this reason, the motives of activism and the criteria of affiliation that guide these organizations, the channels of influence and the lobbying they implement in support of their own interests, and the way they build alliances have only been studied in a rudimentary manner.

Also worth noting is the fact that the rights and demands of the organizations concerned are expressed not so much in the name of general health care and universal rights as on the basis of a particular genetic profile that, by and large, is shared only by a few. This complicates the political articulation of rights, since the accent is placed more on genetic difference than on a common biological identity (Heath, Rapp, and Taussig 2004, 157–159).

Ethopolitics

One of the most influential reworkings of the term "biopolitics" comes from Nikolas Rose. Like Paul Rabinow, with whom Rose works closely (cf., for example, Rabinow and Rose 2006), Donna Haraway (1997) and Hans-Jörg Rheinberger (2000), Rose proceeds on the assumption that the growth of biological and genetic knowledge and the technological practices that emerge from them dissolve the traditional boundary between nature and culture, as well as that between biology and society. In this way, recourse to a pre- or extrapolitical nature is blocked, and biology cannot be separated from political and moral questions. The result of this synthesis, writes Rose, is a new constellation he calls "ethopolitics."

Ethopolitics signifies first of all an epochal rupture. Rose argues that genetics today has little to do with the eugenic interventions of the past. He dismisses critical analyses that view contemporary human genetics as an extension or an intensification of traditional forms of selection and population regulation. Rose holds, on the contrary, that the paradigm of state-enforced policies of extermination and screening is misleading, since the biopolitical frame of reference, as well as biopolitical forms of regulation, have changed. In contrast to "racial hygiene," human genetics today is directed not at the body of the population but at the genetic makeup of the individual. The central goal of genetic interventions, Rose believes, is less the health of the public at large or some other collective idea and more an attempt to improve the health of individuals and to help them avoid illness. In place of state-enacted eugenic programs that usually resorted to repressive methods—from forced sterilization to genocide—we find "a variety of strategies that try to identify, treat, manage, or administer those individuals, groups, or localities where risk is seen to be high" (2001, 7).

These "mutations" of biopolitical rationalities imply a broadening of the scope of biopolitical matters (cf. Rose 2007, 5–7). Corrective and preventive measures no longer target specific, limited subpopulations. All members of society are affected to the extent that everyone is predisposed to genetic risk. The risk discourse here includes those who are currently healthy and submits them to the same medical monitoring as the sick in order to anticipate and—when possible—prevent future illnesses. This expansion of medical territory is part of a general tendency that Rose understands as the "democratization of biopolitics" (ibid., 17). According to Rose, in the 20th century the popularization and adoption of hygienic norms and political measures that targeted health improvement increasingly resulted in individuals taking the initiative when it comes to fighting illness. The dismantling of socialized forms of regulation, along with the establishment of neoliberal programs and policies in the past thirty years, has played an important role in making autonomy and self-determination key elements in medical decision-making (ibid, 3–4).

Rose argues that this coevolution of political transformation and technoscientific innovation has been responsible for a fundamental shift in biopolitical mechanisms. The means of intervention available today, he writes, impinge not only on the appearance and behavior of the body but also on its organic substance, which is now perceived as malleable, correctable, and improvable. In this changed constellation, the body is more and more important for individual identity and self-perception. To the extent that the boundaries between the normal and the pathological, and between healing and enhancement, are increasingly disappearing, a new set of ethical and political questions is emerging that supersedes old-style biopolitics. Rose understands ethopolitics as follows:

> ways in which the ethos of human existence—the sentiments, moral nature or guiding beliefs of persons, groups, or institutions— have come to provide the "medium" within which the self-government of the autonomous individual can be connected up with the imperatives of good government . . . If discipline individualizes and normalizes, and biopower collectivizes and socializes, ethopolitics concerns itself with the self-techniques by which human beings

should judge themselves and act upon themselves to make themselves better than they are. (2001, 18)

The special feature of this form of politics lies in a vital constructivism, which distances itself from ideas of an original, immediately accessible nature and essentialist concepts of human existence. Rose is aware of the ambivalence of this "vital politics" (2001, 22; 2007, 8). On the one hand, the antinaturalist position calls for deep ethical reflection that includes concerns about biological constitution, as well as concepts of identity and how one wishes to live. Individuals can (and must) weigh the various options in order to creatively make the best of transformative possibilities. This politics is about creating individual and collective potentials in a field that was hitherto perceived as immutable. The result could be a pluralization and diversification of the norms of life and of health, which could be opened up for democratic negotiations and decision-making.

On the other hand, the newly won spaces of freedom threaten to revert to their opposites. To begin with, this applies to the commercialization of life processes, which puts research in thrall to the profit motive, and to the development of new forms of social inequality and exploitation (2007, 31–39). Furthermore, Rose sees in the context of ethopolitics the development of new institutional expectations and social norms that point to a "genetic responsibility." A range of "pastoral powers" and authorities crystallize around ethopolitical problems and offer answers to questions regarding the meaning and value of life. Physicians, bioethicists, genetic counselors, scientists, and representatives of pharmaceutical enterprises and biotech companies popularize scientific knowledge, disseminate value judgments, and guide moral reflection (ibid., 40, 73–76). Personal striving for health and wellness is in this way closely allied with political, scientific, medical, and economic interests.

Rose's work is characterized by an impressive dialogue between empirical analysis and theoretical reflection. His texts are among the most frequently cited and exciting in recent sociology. Nonetheless, there are at least two objections to his idea of ethopolitics. The first critical remark concerns the assumption of a clear and distinct rupture between eugenic programs of the past and contemporary human genetic practices. Lene Koch (2004), for example, shows that processes of exclusion and selection in the context of genetic and reproductive technologies cannot be seen as belonging to the past; rather, the forms of intervention and modes of justification have changed. The fundamental objective of controlling and guiding reproductive decisions remains intact. Although it is certainly necessary to stress the historical differences, it is equally important not to erase continuities between past and present.

Second, it remains unclear to what extent biopolitics merges with ethopolitics. Bruce Braun (2007) has drawn attention to the fact that ethopolitics and the ethical questions it addresses are bound to material conditions of life that are unavailable to millions around the world who must fight every day to survive. Yet even if one limits the ethopolitical problematic to Western industrialized states, a central dimension of contemporary biopolitical practices is still missing. To illustrate this, Braun points to the political and media reaction to the spread of avian flu in 2005. He shows that the idea of an isolated and stable molecular body, which for Rose provides the foundation for ethical decisions and practices of the self, can be counteracted through other perceptions of the body. In epidemiological and political discourses regarding the prevention of a given pathogen's spread, an open and vulnerable molecular body is at issue—a body that interacts with other human and nonhuman bodies and is permanently threatened by the risk of disease. A set of political techniques is meant to respond to these dangers, which Braun describes as "biosecurity." Biosecurity aims to guide biological life and its developmental cycles and contingencies. Braun's argument is, in short, that every complete portrait of contemporary biopolitics must embrace issues of biosecurity as well as ethopolitical mechanisms.[2]

8

Vital Politics and Bioeconomy

From *Menschenökonomie* to Human Capital

The concept of vital politics, which Nikolas Rose employs in his discussion of the molecularization and informatization of life, was already in use much earlier in a completely different context. The term played a prominent role in the work of Wilhelm Röpke and Alexander Rüstow, two significant representatives of postwar German liberalism and architects of the social market economy (*soziale Marktwirtschaft*). In the 1950s and '60s, they used the term "vital politics" to refer to a new form of the political that was grounded in anthropological needs and that has an ethical orientation. The negative point of reference here is a mass society that erodes social integration and cohesion. "Massification" (*Vermassung*) is the antonym of vital politics, representing the "worst social malady of our time" (Rüstow 1957, 215). Whereas massification emerged from the dissolution of original social bonds and forms of life, vital politics aims to promote and reactivate them. Contrary to social policy, which focuses on material interests, vital politics takes into account "all factors upon which happiness, well-being, and satisfaction in reality depend" (Rüstow 1955, 70).

The ordoliberal[1] concept of vital politics was the result of a double-pronged approach. According to Rüstow, both the market economies of the West and the socialist states of the East were on the wrong track. Both social systems were in the grip of centralization and were dominated by material concerns. Rüstow wished to reactivate a "natural" principle of politics which, in his view, had

progressively declined since the 19th century. He argued that normative guidelines of political action must consider how policies "affect
well-being and the self-esteem of individuals" (1957, 235). Politics
should resonate with human nature, instead of alienating itself from
it. The yardstick by which this politics is measured is natural and inborn human needs, an orientation which discloses the anthropological foundations of vital politics (ibid., 236).

Politics must adapt to the "essence of the human" (ibid., 235),
which indicates the primacy of politics over the realm of the economic. According to Rüstow, vital politics is founded on the basic
difference between the "good life" and material affluence; it understands the economic system as an integral part of a higher order that
defines and limits the scope of economic activity. Vital politics en-
lists mechanisms of economic coordination and regulation to "serve
life," so that economic measures represent a means to an end rather
than an end in itself.

In Rüstow's view, vital politics is by no means limited solely to a
state's activity but is rather "politics in the widest possible sense. . . .
[I]t encompasses all social measures and experimental arrangements" (ibid., 235). It reactivates moral values and cultural traditions,
while focusing on spiritual solidarity and relationships developed
over time. The goal of this policy is to insert an "ever more dense net
and weave of living ties [*lebendiger Bindungen*] into the entire social
realm" (ibid., 238). This is a task involving both innovation and integration and takes into account all social elements and strata, while at
the same time recognizing their self-organizational capacities. In this
respect, vital politics follows the principle of subsidiarity, because
when it comes to social problems the first concern is whether they
can be solved by autonomous life forms, that is, whether solutions to
a given problem can be found within the sphere of family, neighbor-
hood, and the like, before the state is asked for help (ibid., 232). Rüstow contends that a successful policy depends on families acting as
"basic cells of the social body" and remaining healthy, on "corporate

solidarity" in the workplace, and on the legislative and executive branches of government working for the "integration of the people's body [*Volkskörper*]" entrusted to it (ibid., 237).

Vital politics fulfills two important functions in ordoliberal thinking. First, it serves as a critical principle against which political activity can be measured and which relates the economy back to a comprehensive order that is external to it and ethically grounded. Second, the vital-political dimension of the social market economy asserts its superiority over the "inhumane conditions" existing in the Soviet Union, where fundamental human needs were ignored (ibid, 238).

Whereas for the ordoliberals vital politics points to the conflictual relationship between economic principles and an ethically superior and anthropologically grounded order, there are two 20th-century theories which, by identifying the human being as *homo economicus,* defuse possible conflicts between politics, ethics, and economy. These two theories, the concept of *Menschenökonomie* (human economy) and human capital theory, have less to do with accommodating the economy to life processes than with improving, enhancing, and optimizing those processes. In both cases, human life does not serve as a measure of the economy but is itself subordinated to the economic imperative of valorization.[2]

The concept of *Menschenökonomie* derives from the Austrian social philosopher and sociologist of finance Rudolf Goldscheid, one of the founding members of the German Society for Sociology. His treatise on social biology (1911) sought to provide a comprehensive account of and guide to the management of the conditions of the (re)production of human life. The money the state spends on upbringing, education, and subsistence is contrasted with the profits that human labor generates. The goal of this human-economic calculation is to reach the highest possible "surplus value," that is, to maximize advantages by minimizing expenditures. This "vital optimum" (ibid., 499) requires orderly accounting and allows for an efficient and rational administration and control of "organic capital"—that is, of human labor and life.

Goldscheid distinguishes his thinking from two competing models of social regulation and governance that were widely discussed at the time: social Darwinism and racial hygiene. In his view, these models were not up to the task of optimization. The economic handling of "human material" (1912, 22) cannot be ensured through social Darwinist solutions or racial hygiene experiments. Goldscheid, who was politically allied to the Social Democrats, placed less value on natural or social selection than on the improvement of living conditions, the promotion of education, and the battle against the causes of disease. These efforts targeted the improvement of "human quality" as a whole. Goldscheid states that what "can be observed in any economy" repeats itself at the human level:

> The more carefully an object is made, the higher the expenditure of human labor is required in its manufacture, the more capable and durable will it be. The more expensive, solidly crafted man is the one who grows out of a healthy native soil, who is procreated by healthy fathers. The adolescent person should at least be given the same amount of care and nurture that is performed in the breeding of animals. (1911, 495)

With this point, the idea of a *Menschenökonomie* presents a specific critique of capitalism. Capitalism is guilty of exploiting organic capital, since it deems the satisfaction of human needs irrelevant and does not concern itself with the production of "organic surplus value." In contrast to this, Goldscheid welcomes the socialist alternative of a comprehensively planned economy which would promote the foundation for a rational cultivation of life. From this he anticipates a "restocking of the entirety of the nation's human material" (ibid., 577). Goldscheid understood himself as a humanist. His indictment of the waste of human material resulted in his appeal to conceive of human life as economic capital, so that it would be treated with care and protected from the excesses of capitalist exploitation.

Goldscheid's ideas were grounded in an optimistic belief in social progress and the historical enhancement of the human species. Both of these were to be brought about through the improvement of individual and collective living conditions. However, conceiving of human beings as economic goods might give rise to an entirely different sort of cost-benefit analysis. The economy of solidarity one finds in Goldscheid's writings, for example, was soon displaced by a murderous and selective logic that dispassionately weighed the costs and benefits, expenses and potential revenue, of individuals, as it ranked their relative worthiness to live. After World War I, for example, the lawyer Karl Binding and the physician Alfred Hoche called for "permission to exterminate life unworthy of being lived" (Binding and Hoche 1920). People with disabilities who needed constant care should be killed with impunity, they claimed, a demand that was fulfilled, at the latest, in the murderous Nazi "euthanasia" program, which exterminated people with mental disabilities.

After World War II, human capital theory incorporated Goldscheid's insights without actually referring to them explicitly. Its most prominent representatives were the economists Theodore W. Schultz and Gary S. Becker, who joined with Goldscheid—by then an almost forgotten sociologist from the turn of the century—in a call to "invest in people" (Schultz 1981). Nonetheless, the way to improve the quality of a given population, as Schultz and Becker describe it, is markedly different from Goldscheid's *Menschenökonomie*. Human capital theory breaks with the directed control of a planned economy and installs in its place the indirect effects of the "invisible hand" of spontaneous market regulation. If for Goldscheid the market was still deficient with respect to the targeted accumulation of "organic capital," human capital theory views the market as an unavoidable control instrument used to raise the individual and the collective quality of life. Though the classic works of human capital theory are already a few decades old, the significance of the concept has actually grown since it was formulated, and it has been taken up in the media, in politics, and in everyday communication.

Through the lens of human capital theory, a human being is a rational actor who is constantly allocating scarce resources in the pursuit of competing goals. All activity is presented as a choice between attractive and less attractive alternatives. The basis of this theory is a methodological individualism, whereby a person maximizes benefits and weighs options in a marketplace in which offers and demands coexist in perpetual interplay.

Becker and Schultz understand human capital to mean the abilities, skills, and health, as well as such qualities as the outer appearance and social prestige, of a person. It consists of two components: an inborn corporeal and genetic endowment, and the entirety of the abilities that are the result of "investments" in appropriate stimuli—nutrition, upbringing, and education, as well as love and care. Schultz and Becker write that this "human capital" can be seen as a scarce resource whose restoration, preservation, and accumulation require investment. According to this theory, decisions for or against marriage, for or against having children, or for or against a given career would be interpreted and analyzed as functions of selective choices and preference structures. Thus, men and women marry if they believe this decision to be beneficial, and they file for divorce if this action promises an increase in well-being. Even the desire to have children follows an economic calculus. Children are seen either as a source of psychic pleasure or as future labor that will one day bring in money. Whether it is the desire for children, education, career, or marriage, the claims of this theoretical perspective know no natural limits and extend across the totality of human behavior. The "economic approach" (Becker 1976) conceives of all people as autonomous managers of themselves, who make investment decisions relevant to themselves only and who aim for the production of surplus value. The flipside of this is that they are also responsible for their own failure in the face of social competition, which provides an interesting contrast to the ideas presented by Goldscheid and others in the early 20th century.

In the program of the *Menschenökonomie*, the state sovereign functioned as an idealized global capitalist who strove to accumulate organic surplus value. As Ulrich Bröckling notes, in the years after World War I the state also decided what life was deemed "not worthy of living" and could therefore legitimately be killed (2003, 20–21). With human capital theory, which emerged after World War II, every individual becomes not only a capitalist but also the sovereign of him- or herself. With every action, he or she maximizes his or her individual advantage, but he or she also—to use Foucault's formulation—exerts power in order to "make life or let die." Following the economic approach, diseases and (premature) death could be interpreted as the result of (wrong) investment decisions: "*most* (if not all!) deaths are to some extent 'suicides' in the sense that they could have been postponed if more resources had been invested in prolonging life" (Becker 1976, 10; emphasis in original).

Biocapital

Whereas the concept of *Menschenökonomie* and human capital theory view human existence from a perspective of economic rationality, in recent times a range of political initiatives have postulated that the boundaries and the substance of the economic have to be redefined. The economy, according to this ambitious projection, will soon transform itself into a "bioeconomy."[3] In 2006, the Organization for Economic Cooperation and Development (OECD) published *The Bioeconomy to 2030: Designing a Policy Agenda.* "Bioeconomy" is defined in this programmatic text as a society's sum total of economic operations which use the potential value of biological products and processes in order to create new growth and prosperity for citizens and nations (OECD 2006, 3).

At approximately the same time as the OECD document appeared, the European Commission adopted a plan with a similar goal. The Commission stressed the potential of a "knowledge-based bioeconomy" (KBBE) that would both strengthen European competitiveness in international markets and help to protect the environment. The European Commissioner for Science and Research, Janez Potočnik, described the project as follows: "As citizens of planet Earth it is not surprising that we both turn to 'Mother Earth'—to life itself—to help our economies to develop in a way which should not just enhance our quality of life, but also maintain it for future generations" (European Commission 2005, 2).

Both the European Commission's and the OECD's programs are meant to promote new products and services derived from bioscientific innovations. Central to this vision, therefore, is the creation and regulation of markets rather than a fundamental realignment of the economy, which is implied in the term "bioeconomy." This enlarged meaning of the word appears in scientific works, which in contrast to the political programs observe a decisive and structural transformation of economic relations.

In *Tissue Economies: Blood, Organs, and Cell Lines in Late Capitalism* (2006), a book by the medical anthropologist Catherine Waldby and the literary scholar Robert Mitchell, terms such as "biovalue" or "tissue economies" do not refer to a political economy of capitalist accumulation. These are situated instead in a symbolic economy of gift exchange. On the one hand, blood and other bodily substances are commonly understood as "gifts" that are unselfishly donated to help a needy third party. On the other hand, however, biomaterials are increasingly viewed as commodities that can be sold and traded for profit. Through numerous case studies, the book reveals the limits of a dichotomous and exclusive juxtaposition of gift and commodity exchange and of social and economic logic. These binary models are no longer suited to describing the complex systems of generation, circulation, and acquisition of corporeal materials.

A good insight into the relationship between bioscientific innovations and transformations in capitalism is provided by the anthropologist Kaushik Sunder Rajan in *Biocapital: The Constitution of Postgenomic Life* (2006). Beginning with the findings of science and technology studies that "science" and "society" are not two separate systems or spheres but rather mutually constitutive ones, Sunder Rajan investigates the coproduction of bioscientific knowledge and politico-economic regimes. His empirical thesis is that the emergence of the biosciences marked a new form and a new phase of capitalism (ibid., 3). "Biotechnology" and a genetic understanding of illness are only comprehensible in the light of the capitalist economy's global production and consumption networks. From a theoretical standpoint, Sunder Rajan links Foucault's concept of biopolitics to Marx's critique of political economy, situating both within his anthropological analysis (ibid., 3–15, 78–79). The constitution of biocapital can in turn be mapped through a dual perspective:

[O]n the one hand, what forms of alienation, exploitation, and divestiture are necessary for a "culture of biotechnology innovation" to take root? On the other hand, how are individual and collective subjectivities and citizenships both shaped and conscripted by these technologies that concern "life itself"? (Ibid., 78)

Sunder Rajan's book is based on a multiplicity of field studies, observations, and interviews with scientists, physicians, entrepreneurs, and government representatives in the United States and India. It combines detailed ethnographic research with comprehensive theoretical reflection. Although the book's subject matter is broad, the empirical focus of its analysis is centered on the development of pharmaceuticals, especially the question of how genomic research has transformed their production. An important aspect of contemporary pharmaceutical research aims to create "personalized medicine," that is, medicine whose production is based on the genetic traits of the patient, sometimes known as pharmacogenomics.

Sunder Rajan shows how the scientific production of knowledge can no longer be separated from capitalist production of value. Two risk discourses permeate each other in this area of pharmaceutical research: the medical risk that current and future patients have of facing a major illness, and the financial risk of pharmaceutical companies whose great investment in research and development should ultimately result in commodities. Sunder Rajan describes this branch of industry as a special form of capitalism—a speculative capitalism that is based less on the manufacture of concrete products than on hopes and expectations. It brings together into an "organic" synthesis the hope of patients that new medical treatments will be developed with the zeal of risk capitalism for future profits.

The "new face of capitalism" (ibid., 3) has, in fact, a familiar face. As Sunder Rajan shows through the example of a research hospital in Mumbai, "biocapitalism" reproduces and renews traditional forms of exploitation and inequality. At this hospital, a private company conducts pharmacogenomic studies for Western pharmaceutical companies. Owing to the low cost of working in India and its genetic diversity, it is an especially attractive place for such research. The work takes place in a part of Mumbai that is composed mostly of people who are either poor or unemployed because of the decline of the textile industry. Most of the research subjects have hardly any choice but to participate, "of their own free will," for very little remuneration in clinical studies and to offer their bodies as experimental fields for biomedical study. Despite doing so, they are rarely able to take advantage of the new therapies that might result from such research. Sunder Rajan convincingly shows how global research and clinical studies rely on local conditions and how in "biocapitalism" the improvement or prolongation of one person's life is often linked to the deterioration of the health and the systematic corporeal exploitation of someone else's (ibid., 93–97).

The sociologist Melinda Cooper also studies the relationship between capitalist restructuring and bioscientific innovation from a Marxist perspective. In her book *Life as Surplus: Biotechnology and Capitalism in the Neoliberal Era* (2008), she traces the emergence of an independent biotech industry in the United States at the beginning of the 1970s. At the start of that decade, the Fordist model of accumulation was in decline. It had been based on the coordination of mass production with mass consumption, and it ensured stable growth after World War II. The economic crisis was soon complemented by a growing sensitivity to ecological problems. *The Limits to Growth* (Meadows et al. 1972) and other reports on the environment were powerful reminders not only that the world's resources were limited but also that the effects of industrial production on the climate and the ecosystem were potentially disastrous. According to Cooper, the promise of a "bioeconomy" was an answer to this dual crisis. She suggests that the "biotech revolution" can be conceived of as part of a comprehensive "neoliberal revolution" and its attempt to restructure the U.S. economy: "Neoliberalism and the biotech industry share a common ambition to overcome the ecological and economic limits to growth associated with the end of industrial production, through a speculative reinvention of the future" (2008, 11).

Cooper's book investigates various aspects and dimensions of this "neoliberal biopolitics" (ibid., 13). Taking up Foucault's analysis in *The Order of Things* (1970) of the mutual constitution and permeation of biology and political economy, she proceeds from the assumption that biological processes are increasingly entangled with capitalist strategies of accumulation and are becoming a new source of surplus value generation. At the same time, however, life processes are not simply becoming a new object of exploitation and expropriation. Rather, neoliberal capitalism is itself adopting a "biological" format. It "lives" from the vision of biological growth that can overcome all natural limits.

Cooper's often rather speculative but always intriguing analysis links the debt-driven growth of the United States and its exorbitant deficit with NASA's astrobiological research on extraterrestrial forms of life, which at some point—it is hoped—will overcome the limitations of living on Earth. She also examines, on the one hand, the transfer of ideas between theoretical biology and its deliberations over evolution and complex life processes and, on the other, the neoliberal rhetoric of limitless economic growth, which in recent times has drawn on vitalist concepts. Both bodies of thought stress the potential of self-organization and criticize models of equilibrium. They celebrate crises of developmental processes as fertile ground for dynamic innovation and adaptation that are supposed to transcend existing economic or natural limitations. *Life as Surplus* shows in vivid detail the interconnections and correspondences of apparently unrelated discourses and practices and is itself the result of a synthesis. The book makes plausible the argument that an analysis of biopolitics cannot be separated from a critique of the political economy of life.[4]

However, the works introduced in this chapter are still exceptions. On the whole, only very few studies that employ the term "biopolitics" have pursued the question of how the politicization of life is intertwined with its economization.

9

Prospect: An Analytics of Biopolitics

THE OVERVIEW OF the history and contemporary uses of "biopolitics" presented in this book reveals that the term is a combination of apparently contradictory elements. If politics in the classical sense refers to a state beyond existential necessities, biopolitics introduces a reflexive dimension. That is to say, it places at the innermost core of politics that which usually lies at its limits, namely, the body and life. Seen this way, biopolitics again includes the excluded other of politics. Indeed, neither politics nor life is what it was before the advent of biopolitics. Life has ceased to be the assumed but seldom explicitly identified counterpart of politics. It is no longer confined to the singularity of concrete existence but has become an abstraction, an object of scientific knowledge, administrative concern, and technical improvement.

And politics? Politics has also changed in the light of biopolitical rationalities and technologies. It has made itself dependent on life processes that it cannot regulate and whose capacities for self-regulation it must respect. However, it is precisely this limitation that has provided politics with many options for different forms of intervention and organization. Politics disposes not only of direct forms of authoritative command but also of indirect mechanisms for inciting and directing, preventing and predicting, moralizing and normalizing. Politics can prescribe and prohibit, but it can also incite and initiate, discipline and supervise, or activate and animate.

Let us look again at the naturalist and politicist conceptions of biopolitics described earlier in this book. Both fundamental positions appear to be constitutive elements of a common biopolitical problematic. The conception of nature as deterministic and shackled to its fate is the flipside of its increasing permeation by science and technology (cf. Latour 1993). Both perspectives diminish the significance of politics by conceiving of it as reactive, deductive, and retroactive. The naturalist interpretation limits itself to reproducing the order of nature and expressing what has been predetermined by biological processes. In the politicist variant, politics seems to be merely a reflex of scientific and technological processes because it only regulates how society adapts to these developments.

To counter these two fundamental, and at the same time contrasting and complementary, positions, I here sketch an analytics of biopolitics that takes the significance of the political seriously. This perspective distinguishes itself from naturalist and politicist conceptions because it focuses on neither the causes nor the effects of the politics of life, describing instead its mode of functioning. At its center we find the question "how?" instead of "why?" or "what for?" It deals with neither the biologization of politics nor the politicization of biology, since "life" and "politics" are conceived of as elements of a dynamic relationship, rather than as external and independent entities.

This analytics of biopolitics has its starting point in the theoretical perspective outlined by Michel Foucault, but it "lives," so to speak, from the numerous corrections and elaborations of biopolitics that are at the core of this book. Taken together, these lines of reception have advanced and substantiated the Foucauldian notion of biopolitics in different ways. First, they make clear that contemporary biopolitical processes are based on an altered and expanded knowledge of the body and biological processes. Thus, the body is conceived of as an informational network rather than a physical substrate or an anatomical machine. Second, it was necessary to supplement the analysis of biopolitical mechanisms with an examination of the modes of subjectivation. This theoretical move allows us to assess how the regulation of life processes affects individual and collective actors and gives rise to new forms of identity. In short, following Foucault, recent studies of biopolitical processes have focused on the importance of knowledge production and forms of subjectivation. An analytics of biopolitics should investigate the network of relations among power processes, knowledge practices, and modes of subjectivation. Accordingly, it is possible to distinguish three dimensions of this research perspective (see also Rabinow and Rose 2006, 197–198).[1]

First, biopolitics requires a systematic knowledge of "life" and of "living beings." Systems of knowledge provide cognitive and normative maps that open up biopolitical spaces and define both subjects and objects of intervention. They make the reality of life conceivable and calculable in such a way that it can be shaped and transformed. Thus, it is necessary to comprehend the regime of truth (and its selectivity) that constitutes the background of biopolitical practices. One must ask what knowledge of the body and life processes is assumed to be socially relevant and, by contrast, what alternative interpretations are devalued or marginalized. What scientific experts and disciplines have legitimate authority to tell the truth about life, health, or a given population? In what vocabulary are processes of life described, measured, evaluated, and criticized? What cognitive and intellectual instruments and technological procedures stand ready to produce truth? What proposals and definitions of problems and objectives regarding processes of life are given social recognition?

Second, as the problem of the regime of truth cannot be separated from that of power, the question arises of how strategies of power mobilize knowledge of life and how processes of power generate and disseminate forms of knowledge. This perspective enables us to take into account structures of inequality, hierarchies of value, and asymmetries that are (re)produced by biopolitical practices. What forms of life are regarded as socially valuable, and which are considered "not worth living"? What existential hardships, what physical and psychic suffering, attract political, medical, scientific, and social attention and are regarded as intolerable and as a priority for research and in need of therapy, and which are neglected or ignored? How are forms of domination, mechanisms of exclusion, and the experience of racism and sexism inscribed into the body, and how do they alter it in terms of its physical appearance, state of health, and life expectancy? Also, this perspective investigates the "economy" of the politics of life: who profits and how from the regulation and improvement of life processes (in terms of, for example, financial gain, political influence, scientific reputation, and social prestige)? Who bears the costs and suffers such burdens as poverty, illness, and premature death because of these processes? What forms of exploitation and commercialization of human and nonhuman life can be observed?

Third, an analytics of biopolitics must also take into account forms of subjectivation, that is, the manner in which subjects are brought to work on themselves, guided by scientific, medical, moral, religious, and other authorities and on the basis of socially accepted arrangements of bodies and sexes. Here again one can formulate a complex of questions that presents some relevant issues: How are people called on, in the name of (individual and collective) life and health (one's own health and that of the family, nation, "race," and so forth), in view of defined goals (health improvement, life extension, higher quality of life, amelioration of the gene pool, population increase, and so forth) to act in a certain way (in extreme cases even to die for such goals)? How are they brought to experience their life as "worthy" or "not worthy" of being lived? How are they interpellated as members of a "higher" or "inferior" race, a "strong" or a "weak" sex, a "rising" or a "degenerate" people? How do subjects adopt and modify scientific interpretations of life for their own conduct and conceive of themselves as organisms regulated by genes, as neurobiological machines, as composed bodies whose organic parts are, in principle, exchangeable? How can this process be viewed as an active appropriation and *not* as passive acceptance?

What does such an approach contribute to the understanding of contemporary societies? Where is its "theoretical surplus value" to be found (cf. Fassin 2004, 178–179)? From a *historical* perspective, an analytics of biopolitics demonstrates not only how, in the past few centuries, the importance of "life" for politics has increased but also how the definition of politics itself has thereby been transformed. From reproductive cloning to avian flu to asylum policies, from health provision to pension policies to population declines—individual and collective life, their improvement and prolongation, their protection against varieties of danger and risk, have come to occupy more and more space in political debate. Whereas the welfare state has been able until recently to focus on the problem of securing the lives of its citizens, today the state also has to define and regulate the beginning and the end of life. Thus, the question of who is a member of the legal community or, to put it another way, the question of who is not yet or no longer a member of the legal community (embryos, the brain dead, etc.) becomes acute.

In *empirical* terms, an analytics of biopolitics can bring together domains that are usually separated by administrative, disciplinary, and cognitive boundaries. The categorical divisions between the natural and the social sciences, body and mind, nature and culture lead to a blind alley in biopolitical issues. The interactions between life and politics cannot be dealt with using social-scientific methods and research models alone. The analysis of biopolitical problems necessitates a transdisciplinary dialogue among different cultures of knowledge, modes of analysis, and explanatory competences. Similarly, it is inadequate to isolate the medical, political, social, and scientific aspects of biopolitical questions from one another. The challenge of an analytics of biopolitics consists precisely in presenting it as part of a greater context—a context that contains numerous divisions in the form of empirical facts that could be explained historically and perhaps overcome or at least shifted in the future.

Finally, an analytics of biopolitics also fulfills a *critical* function. It shows that biopolitical phenomena are not the result of anthropologically rooted drives, evolutionary laws, or universal political constraints. Rather, they have to be grounded in social practice and political decision-making. These processes do not follow a necessary logic but are subject to specific and contingent rationalities and incorporate institutional preferences and normative choices. The task of an analytics of biopolitics is to reveal and make tangible the restrictions and contingencies, the demands and constraints, that impinge upon it.

The critical aspect here does not consist in a rejection of what exists. Rather, it seeks to generate forms of engagement and analysis that enable us to perceive new possibilities and perspectives or to examine those that already exist from a different point of view. Critique here is productive and transformative, rather than negative or destructive. It does not promise to deliver an ultimate and objective representation of reality based on universal claims of scientific knowledge; on the contrary, it critically assesses its own claims and exposes its own particularity, partiality, and selectivity. Instead of being grounded on authoritative knowledge, an analytics of biopolitics has an ethicopolitical orientation: an "ethos" or a "critical ontology of ourselves" (Foucault 1997b, 319).² This critical ethos allows a path to be forged that leads beyond the futile choice between the trivialization and the dramatization of biopolitical phenomena. It is not convincing to deproblematize biopolitics for the purpose of presenting it as a seamless extension and amplification of millennia of agricultural methods of production and breeding, as Volker Gerhardt does. Nor does it make sense to exaggerate the issue and suggest, as Agamben does, that Auschwitz is the apogee of biopolitics. Although these two positions present opposing accounts of biopolitics, they nevertheless both prioritize some general normative preferences instead of offering an empirical analysis.

The critical ethos of an analytics of biopolitics might also disrupt the current institutional and discursive dominance of bioethics.

Bioethics has narrowed the terms of public debate on the relations between life and politics, since the discussion is mainly conducted in ethical terms and as an argument about values (cf. Gehring 2006, 8–9; Wehling 2007). Whereas an analytics of biopolitics offers us a way of perceiving the complexity of a relational network, bioethical discourse obscures the historical genesis and social context of biotechnological and biomedical innovations in order to present alternative options for decision-making. Thus, it fails to account for the epistemological and technological foundations of life processes and their integration into power strategies and processes of subjectivation. The emphasis in bioethics is on abstract choices, and there is no examination of who possesses (and to what degree) the material and intellectual resources actually to use specific technological or medical options. Also, bioethics often neglects the social constraints and institutional expectations that individuals might experience when they wish to take advantage of the options that, in principle, are available to them.

Bioethics focuses on the question, what is to be done? It reduces problems to alternatives that can be treated and decided. It gives answers to specific demands. An analytics of biopolitics, on the other hand, seeks to generate problems. It is interested in questions that have not yet been asked. It raises awareness of all those historical and systematic correlations that regularly remain outside the bioethical framework and its pro-contra debates. An analytics of biopolitics opens up new horizons for questioning and opportunities for thinking, and it transgresses established disciplinary and political borders. It is a problematizing and creative task that links a diagnostics of the contemporary with an orientation to the future, while at the same time destabilizing apparently natural or self-evident modes of practice and thought—inviting us to live differently. As a result, an analytics of biopolitics has a speculative and experimental dimension: it does not affirm what is but anticipates what could be different.

Notes

Notes to the Introduction

1. Cf. the contributions to the *Lessico di biopolitica* (Encyclopedia of Bio-politics) (Brandimarte et al. 2006).

2. By "politicism" I mean the idea of the political domain as a self-contained and self-reproducing unity, which tends to exaggerate the autonomy of the political (see Jessop 1985, 73).

Notes to Chapter 1

1. For a brief history of the concept of biopolitics, see Esposito 2008, 16–24.

2. The first political scientist to use the concept of biopolitics in this sense was probably Lynton K. Caldwell (1964).

3. In Germany, Heiner Flohr, now emeritus professor of political science at the University of Düsseldorf, has for thirty years argued consistently for the importance of this research perspective (Flohr 1986; cf. Kamps and Watts 1998).

Notes to Chapter 2

1. The *Zeitschrift für Biopolitik* (Journal of Biopolitics), founded in 2002 and since discontinued, contained a number of examples of this interpretive tendency (cf. Mietzsch 2002).

Notes to Chapter 3

1. Foucault here conceives of death in a broad sense, which extends not only to physical killing but also to all social and political forms of death, which he characterizes as "indirect murder": "exposing someone to death, increasing the risk of death for some people, or, quite simply, political death, expulsion, rejection, and so on" (2003, 256).

2. For a critique of Foucault's analysis of racism, see Stoler 1995 and Forti 2006.

3. For a thorough account of the Foucauldian concept of government, see Lemke 1997.

Notes to Chapter 4

1. A more detailed discussion of Agamben's understanding of biopolitics can be found in Lemke 2007, 89–110.

2. Agamben here draws extensively on Hannah Arendt's thinking as developed in *The Origins of Totalitarianism* regarding the "Perplexities of the Rights of Man" and the production, through the modern nation-state, of people without a state and therefore without any rights (Arendt 1968, 267–302). Cf. also Kathrin Braun's (2007) instructive comparison of the biopolitical concepts of Foucault and Arendt.

3. For two important investigations into the role of death in the constitution of modern politics as well as in liberal economies, see Mbembe 2003; Montag 2005.

4. Agamben's considerations on this point lead to his concept of "form-of-life," which "must become the guiding concept and the unitary center of the coming politics" (2000, 12). He understands this as a life that never can be separated from its form, a life in which it is never possible to isolate bare life.

Notes to Chapter 5

1. The first issue of the journal *Multitudes* had the title "Biopolitique et biopouvoir" (2000). Cf. also Lazzarato 2000; Revel 2002; Virno 2004.

Notes to Chapter 6

1. For a further discussion of this problem, see Butler 1998; Fraser and Honneth 2003.

2. See Esposito 1998, 2002. For the place of Esposito and his concept of biopolitics within contemporary philosophy, see Campbell 2008 and the contributions to the special issue of *Diacritics* "Bíos, Immunity, Life: The Thought of Robert Esposito," 36 (2) (2006).

Notes to Chapter 7

1. Today, Rabinow endorses a more cautious view, stressing the limits of the concept he originally formulated in the "Golden Age of molecular biosociality": "There was hope, there was progress, there was a reason to be urgent even strident—there were reasons to want to be biosocial" (2008, 190). However, as Rabinow admits, up until now only very few of the promises of genetic medicine have been realized, and there are hardly any adequate risk-assessment procedures or medical treatments available.

2. On biopolitics and security, see also Reid 2006; Dauphinee and Masters 2007; Dillon and Lobo-Guerrero 2008, as well as the research of the Biopolitics of Security Network: http://www.keele.ac.uk/research/lpj/bos/ (accessed 17 December 2009).

Notes to Chapter 8

1. The term "ordoliberal" is derived from the journal *Ordo,* in which many representatives of German postwar liberalism published.

2. The following account is based largely on the analysis of Ulrich Bröckling, who contrasts these two concepts in one of his essays (Bröckling 2003).

3. A good overview of the complex field of bioeconomics and its various meanings can be found in an already somewhat dated bibliography: Ghiselin 2001 (cf. also *Distinktion* 2007).

4. For another attempt to link Marx and Foucault in an analysis of the contemporary biotech industry, see Thacker 2005.

Notes to Chapter 9

1. The proposal presented here seeks to combine two concepts coined by Foucault, governmentality and biopolitics, in order to conceive of biopolitics as an "art of government" (cf. Lemke 2007).

2. It should be mentioned that this critical ethos displays a range of similarities with, as well as differences to, Adorno's diagnosis of a "damaged" life (cf. Adorno 2006).